A NOTE ON THE AUTHOR

Nicky Jenner is a science writer and editor. Her work has appeared in a variety of international magazines, including *New Scientist, Nature, BBC Sky at Night, Physics World* and *Astronomy Now.* She writes and edits for the European Space Agency, European Southern Observatory (the organisation responsible for both the Very Large Telescope and the upcoming European Extremely Large Telescope), and the Hubble Space Telescope, for which she was formerly the European press officer. Nicky lives in Bristol, UK.

Also available in the Bloomsbury Sigma series:

p53: The Gene that Cracked the Cancer Code by Sue Armstrong
Atoms Under the Floorboards by Chris Woodford
Spirals in Time by Helen Scales
A is for Arsenic by Kathryn Harkup
Breaking the Chains of Gravity by Amy Shira Teitel
Suspicious Minds by Rob Brotherton
Herding Hemingway's Cats by Kat Arney
Electronic Dreams by Tom Lean
Sorting the Beef from the Bull by Richard Evershed and Nicola Temple
Death on Earth by Jules Howard
The Tyrannosaur Chronicles by David Hone
Soccermatics by David Sumpter
Big Data by Timandra Harkness
Goldilocks and the Water Bears by Louisa Preston
Science and the City by Laurie Winkless
Bring Back the King by Helen Pilcher
Furry Logic by Matin Durrani and Liz Kalaugher
Built on Bones by Brenna Hassett
My European Family by Karin Bojs
Patient H69 by Vanessa Potter
Catching Breath by Kathryn Lougheed
PIG/PORK by Pía Spry-Marqués
The Planet Factory by Elizabeth Tasker
Wonders Beyond Numbers by Johnny Ball
Immune by Catherine Carver
I, Mammal by Liam Drew
Reinventing the Wheel by Bronwen and Francis Percival
Making the Monster by Kathryn Harkup
Best Before by Nicola Temple
Catching Stardust by Natalie Starkey
Seeds of Science by Mark Lynas
Eye of the Shoal by Helen Scales
Outnumbered by David Sumpters
Nodding Off by Alice Gregory
The Science of Sin by Jack Lewis

4TH ROCK FROM THE SUN

THE STORY OF MARS

Nicky Jenner

BLOOMSBURY SIGMA
LONDON • OXFORD • NEW YORK • NEW DELHI • SYDNEY

BLOOMSBURY SIGMA
Bloomsbury Publishing Plc
50 Bedford Square, London, WC1B 3DP, UK

First published in the United Kingdom in 2017
This edition published 2018

Photo credits (t = top, b = bottom, l = left, r = right, c = centre)
Colour section: P. 1: NASA / JPL-Caltech / MSSS (c); NASA / JPL-Caltech (b). P. 2: NASA/ JPL (tl); NASA / JPL / Malin Space Science Systems (cl); ESA / DLR / FU Berlin (G. Neukum), MOC (Malin Space Science Systems) (b). P. 3: NASA / JPL (tl); NASA / JPL / Malin Space Science Systems (c); NASA / JPL-Caltech (bl). P. 4: NASA / JPL / University of Arizona (tl, cl) NASA / JPL-Caltech / MSSS (b); NASA / JSC / Stanford University (br). P. 5: Viking Project / NASA / U.S. Geological Survey (USGS) Astrogeology Science Centre (tl, tr); NASA, ESA and Z. Levay (STScl) (b). P. 6: ESA/DLR/ FU Berlin, CC BY-SA 3.0 IGO (creativecommons.org/licenses/by-sa/3.0/igo/) (l); Viking Project / NASA / JPL / U.S. Geological Survey (USGS) Astrogeology Science Centre (cr); NASA / JPL-Caltech / MSSS (br). P. 7: NASA / JPL (t); NASA / JPL-Caltech / University of Arizona (cr); NASA / JPL-Caltech / MSSS / Texas A&M University (br). P. 8: ESA / DLR / FU Berlin, CC By-SA IGO 3.0 (creativecommons.org/licenses/by-sa/3.0/igo/) (t); ESA / DLR / FU Berlin (G. Neukum) (cr); NASA / JPL / USGS (br).

A catalogue record for this book is available from the British Library

Library of Congress Cataloguing-in-Publication data has been applied for

ISBN: PB: 978-1-4729-2252-6; eBook: 978-1-4729-2251-9

2 4 6 8 10 9 7 5 3 1

Typeset by Deanta Global Publishing Services, Chennai, India
Printed and bound in Great Britain by CPI Group (UK) Ltd, Croydon CR0 4YY

To find out more about our authors and books visit www.bloomsbury.com. and sign up for our newsletters

Table of Contents

CHAPTER ONE

Mars Fever

For as long as humans have existed, we've gazed at the sky and dreamt of exploring the Universe. We've stared at the stars and planets nearest to us, noting their dancing motions through the sky and their bright, mottled surfaces, learning more about their characteristics and discovering little families of moons, majestic ring systems, immense swirling storms and volcanoes taller than the very tallest tip of Earth.

Our enthusiasm for exploring the cosmos has been around for centuries. We've sent spacecraft to explore and land on asteroids and comets, and paid visits – some fleeting fly-bys, others longer-lived landers, orbiters and rovers – to every planet in the Solar System. We've sent humans to walk around, drive and play golf on the surface of the Moon, and launched willing human guinea pigs into space to spend months at a time in a cramped tin can falling around the Earth, occasionally suiting up and venturing out into space.

Our fascination with space is understandable. The Solar System alone is packed with fascinating and exciting worlds: the mysterious, scorching surface of smoggy Venus, the delicate rings of the Saturnian system, the diverse and distinctive moons of Jupiter, the heart-stamped surface of distant Pluto (it may no longer be a planet, but it's far from boring). However, one planet has captivated stargazers like no other and continues to do so today: Mars, the Red Planet.

'Mars is a favourite target,' said Dr Firouz Naderi, director for Solar System Exploration at NASA's Jet Propulsion Laboratory in California, US, in a 2003 NASA release preceding the landing of the hugely successful *Spirit* and *Opportunity* rovers. 'We – the United States and former USSR – have been going to Mars for 40 years. The first time we flew by a planet, it was Mars. The first time we orbited a planet, it was Mars. The first time we landed on a planet it was Mars, and the first time we roved around the surface of a planet, it was Mars. We go there often.'

Even if your primary relationship with Mars is in the form of the eponymous chocolate bar – and who would blame you? – it's likely that this single-minded focus over the years has caused you to know at least a little about the planet itself. Mars is the most-googled planet after Earth itself, and we've sent more probes to Mars than any body in the Solar System bar the Moon (or tried to, anyway). Images and data flood in thick and fast from the NASA rovers trundling around on the planet's surface, and from the international fleet of robots currently in orbit around Mars. Numerous governmental and private organisations from all over the world have plans in the pipeline for visiting Mars – manned and unmanned – in the next decade or two, and new announcements seem to pop up weekly.

It's easy to get caught up in the excitement, but it's important to justify our love affair with the Red Planet – especially when missions to Mars are slated to cost a lot of money in the near future and have the added complication of failing unusually often. Why do we want to visit Mars so badly? We have a whole host of intriguing and unique worlds in our little patch of space. Is Mars really so special?

A quick-fire tour of the Red Planet

'Mars needs YOU!' announced a series of light-hearted NASA posters designed in 2009. 'Explorers wanted on the journey to Mars!'

Fancy working a night shift on Mars's largest moon, Phobos? Are you a green-thumbed farmer longing to rake the red-hued soil, harvesting lettuce, radishes, peas and tomatoes for the outpost's canteen? Perhaps you're more of the roving explorer type, and would prefer to survey and map the alien terrain – or you'd be more partial to teaching future explorers all there is to know about the Red Planet and its little cosmic family. Hiking and caving enthusiasts can enjoy the largest canyon in the Solar System (Valles Marineris!), geologists can marvel over the out-of-this-world rock forms, mammoth volcanoes (Olympus Mons!) and chalky dust, and astronomers can track Phobos's speedy backwards race through the pink-tinted sky. Mars has it all – if you could only afford the trip.

With rose-tinted glasses discarded, Mars would actually be an appalling choice of holiday destination for, well, mostly everyone (avid wannabe space explorers, such as the author, excepted).

For one, the trip there would be thumb-twiddlingly boring. Mars is the fourth planet from the Sun after Mercury, Venus and Earth, and is bounded on its outer side by the rocky asteroid belt. The planet sits hundreds of millions of kilometres away from Earth. The exact distance changes depending on where the two planets are sitting in their orbits, but the average separation is around 225 million kilometres (140 million miles).

Even in the best-case scenario, it'd take many months to get to Mars. You're looking at an average trip of eight or so months to travel there and the same to get back. Due to the conditions and planetary alignments needed for launch, you'd also need to spend a few months at Mars before you could even think about heading home. Assuming everything went according to plan, a round trip would take somewhere between 18 months and three years, depending on the type of orbit used – and that's without any time spent on the planet's surface.

The surface itself is freezing, arid, dusty and dry. The entire planet is essentially a huge desert. It's covered in tiny grains of sand and soil and broken-down rock ('regolith') that are rich in rusting iron – painting the planet in its characteristic hues of yellow, orange, red, brown and butterscotch – and a thick coating of dust that resembles finely milled talcum powder in texture. This dust sweeps and wraps around the planet's surface, interfering with any equipment we currently have sitting on the Red Planet and coating it in thick, powdery, perpetual blankets of detritus.

Martian dust storms grow larger than any on Earth. Every year Mars experiences incredible continent-sized dust storms that last for weeks at a time – and every few years these storms grow so large that they cover the entire planet. These storms are thought to be caused by sunlight heating the planet, causing warm air to rise and carry dust with it as it does so. The scale of Mars's storms is likely worsened by the fact that Mars is mostly desert, allowing storms to grow and scoop up material almost endlessly with few obstructions. Dust storms can cover huge portions of Mars's surface for weeks or months on end, whipping up dust devils and throwing dust particles into the Martian sky. Dust often hangs in the air, dimming and tinting it in shades of pinky-red and reducing the amount of sunlight filtering down to the surface.

Mars's atmosphere is predominantly carbon dioxide (over 95 per cent, compared with Earth's 0.04 per cent), contains barely any oxygen (0.13 per cent versus Earth's 21 per cent) and is very tenuous. Any sightseers would be unable to breathe the thin 'air' without special space gear and their own oxygen supply. However, this is a secondary concern; humans would be unable to casually walk around on Mars due to the incredibly low surface pressures. Here on Earth, the weight of all the atmosphere and air sitting above us pushes and presses down on us, putting pressure on our skin. On Earth, 'one atmosphere' is 1,013 millibars at sea level. On Mars there's simply less air pushing downwards, resulting in far lower pressures. The average atmospheric pressure on Mars is 6 millibars – just 0.6 per cent of Earth's!

We need a certain level of surface pressure to keep our bodies pressurised and balanced; as atmospheric pressure lowers, the liquid within our bodies – saliva, sweat, blood, fluid in our eyes and soft tissue – begins to turn into gas, forming bubbles in our vessels and membranes and essentially boiling away. This is known as 'ebullism' and is similar to the cause of 'the bends', a decompression sickness experienced when divers resurface too rapidly. Even if the Martian atmosphere were full of oxygen, our lungs still wouldn't have the sufficient pressure to breathe it, and would either burst, tear or struggle to exhale, leaving our blood and thus organs lacking oxygen (causing 'hypoxia', or oxygen deprivation). Our tissues would swell and bruise and we would bloat – not the best holiday look – and quickly fall unconscious as our brains become starved of oxygen. Luckily, these issues can be fixed by using pressurised spacesuits and oxygen supplies similar to those currently used by astronauts during spacewalks and when walking on the Moon (although innovative and far less restrictive designs are under way).

Mars's lack of atmosphere also means that the planet is constantly bombarded with radiation from both deep space and from the Sun, which releases a continuous stream of charged particles out into the Solar System. Our planet's atmosphere – and, crucially, its magnetic field – acts as a giant cosmic shield, redirecting and deflecting the majority of this radiation away from us. Mars doesn't have that luxury as its magnetic field is almost non-existent, thought to have switched off long ago. Radiation can wreak havoc on electronics systems and badly damage human tissue, breaking down the structure of our cells and increasing the risk of cataracts, radiation sickness, tumours, cancer and more. Until we figure out how to effectively shield against this influx of cosmic radiation on the surface of Mars, it's likely we'd need to spend almost all of our time underground. Again, not an ideal holiday for most.

Coupled with its greater distance from the Sun, Mars's thin atmosphere means that its surface is far colder than Earth's. The planet receives less light (and therefore heat) from the Sun in the first place, but has a harder time keeping hold of it due to its comparably puny atmosphere. Mars has an average temperature of just over -60°C (-76°F), and regularly hits temperatures as low as -90°C (-130°F). When and where things get particularly nippy, this can fall as low as -153°C (-243°F)! Additionally, Mars's thin atmosphere gives it very poor control over its climate, so temperatures fluctuate wildly depending on location, time of day and distance from the surface – at the equator in the summer, for example, they can soar to a pretty temperate 20°C or even 30°C (68–86°F). If you were to stand on the surface of Mars your feet would be noticeably warmer, maybe even tens of degrees warmer, than your head. This isn't impossible for humans to endure, but certainly isn't comfortable.

NASA's *Curiosity* rover found Mars's climate to be not quite as harsh as we expected, so there may be hope yet. The industrious little robot detected temperatures regularly rising above freezing even in the Martian winter, which could be a good sign for future habitability. Here there's a surprising upside to Mars's missing atmosphere: even sub-zero temperatures don't feel quite as bone-chilling as we might expect due to the lack of wind chill. When we envisage cold or desolate locations we often assume there will be a chilling, crisp wind from which we need to wrap up and shield ourselves. However, on Mars, wind speeds max out at around 100kmph (60mph) due to the far thinner air – not gentle, but slower than hurricanes on our planet. Although Mars's dust storms are huge, they're nowhere near as destructive as severe weather storms experienced on Earth. The opening scene of *The Martian* wouldn't have been nearly as exciting without a little artistic licence.

Although it's undeniably an inhospitable and alien world, Mars is also somewhat similar to our planet, especially when compared with the other planets in the Solar System. For one, Mars is terrestrial. Simply put, it has a solid, rocky surface that you could step on to. It has a solid crust formed from rock and metal, wrapped around a small metal core primarily composed of iron, nickel and sulphur. We think this core may be partly molten, partly solid, like Earth's, but we're still unsure. Mars's crust is quite thick – 50km thick on average and very variable, around three times as thick as Earth's relative to diameter – and contains many metals and minerals familiar to us: oxides and silicates, iron, hydrogen, oxygen, aluminium, potassium, calcium, magnesium, clay minerals, organic (carbon-containing) molecules and more.

The planet's surface gravity is low, at around 38 per cent that of Earth's ($3.71 m/s^2$ vs. $9.81 m/s^2$). This value is just over double that of the Moon. Just as the Apollo astronauts leapt and skipped around on the lunar surface, any Martian residents would have to adjust their gait to adapt to this lower gravity. *The Martian* ignored this change in gravity for ease of filming. If they'd included it, the protagonist, Mark Watney, would have been able to leap as high as 2.7m (9ft) into the air and 'walked' with longer steps, assuming more of a loping, bounding, half-hopping stride – not quite as fun as bouncing around on the Moon and not quite as dignified as sauntering around on Earth.

Martian gravity is low because Mars itself is far smaller and less massive than Earth. With a radius of 3,390km (2,106 miles) Mars is just over half the size of our planet, making it the second smallest planet in the Solar System and big sibling only to Mercury. Despite its smaller size, the lack of oceans on Mars means that it has roughly the same amount of dry land area as Earth. Because they're both rocky, we see a lot of the features on Mars that we recognise on our own planet – volcanoes, canyons, craters, gullies, polar caps, features we believe to be old dried-up riverbeds and channels, evidence of erosion, sand dunes, glaciers – but the Red Planet's grow far larger, taller, deeper and more extensive. We also see a lot of sedimentary rock, whose layers, formed via slow compression over many millions of years, can help us to unravel a planet's past. Geologically, we see a lot of ourselves in Mars.

Mars spins on its axis at about the same rate as Earth does, making a Martian day only around 40 minutes longer than a terrestrial one (24 hours, 37 minutes and 23 seconds vs. our 23 hours, 56 minutes and 4.1 seconds)*. A Martian year is also longer than a terrestrial one, but not significantly so on a cosmic scale – it takes Mars 687 days to complete one lap of the Sun, which is just under two Earth years (1.88 to be exact). The planet is tilted on its axis just a couple of degrees more than Earth (25.2° vs. our 23.5°), meaning that the planet also experiences seasons, although these are far more extreme than our own.

---

* This is a 'sidereal day', defined as the amount of time it takes Mars to complete one rotation on its axis (and measured with respect to the apparent motion of the stars). There's also the 'solar day', which is the passage of time based on the Sun's apparent motion through the sky. On Mars a solar day, or *sol*, is 39 minutes and 35 seconds longer than on Earth (which is 24 hours and is the time used in our everyday timekeeping). One *sol* is 1.027 Earth days. A Martian year is 669 *sols*.

The planet has a couple of small, rocky moons, ticking another Earth-like characteristic off the list. Mars's two cosmic companions are named Phobos and Deimos. Unlike our own picture-perfect satellite, the duo look a bit like lumpy, bumpy potatoes and are thought to be little more than piles of space rubble loosely held together. Phobos orbits closer to Mars than any other Solar System moon does to its host planet, whizzing around Mars so fast that it rises and sets multiple times in the sky in a single day. We're still unsure of how these two moons formed – Are they captured asteroids? Were they formed by a massive collision? – and know little about them with any certainty.

When choosing a planet to explore, one key consideration is water (and by extension, life). The Red Planet is dry, but not completely parched. It has polar caps formed partially of water ice, and evidence is growing that there may be stores of water locked up in its soil and beneath its surface. While we once thought Mars to be completely bereft of water, we now know it does have some. We've spotted water in the form of permafrost and brines that seasonally melt and refreeze, streaking their way down crater walls and inclines (features known as 'recurring slope lineae', or RSL), and there's some water vapour in its atmosphere.

But Mars wasn't always so arid. We believe that the planet used to be flooded and soaked with water, its surface covered with rivers and streams and gullies and even vast oceans. The *Curiosity* rover, for example, soon discovered that its landing site was actually an ancient streambed! All of this water somehow slowly left Mars as the planet evolved, leaving behind the waterless world we see today. However dry it now seems, early Mars likely had a thicker atmosphere and an abundance of water – both great signs for potential life.

Seeing red

Mars has risen above the other Solar System planets in terms of appeal. Interest in the planet may be at an all-time high, but in reality we've been fascinated with Mars for centuries, if not longer. It's present everywhere you look and has been for years: astrology, myth and legend, some of the most iconic songs, Hollywood films and older cult classics, childhood cartoons, science fiction and more.

We've been aiming spacecraft at Mars since the 1960s – and have attempted more launches to the planet than to any other body bar the Moon – and have learned a huge amount about the dusty world in the decades since. Mars remains as fascinating to scientists as ever. It offers an ideal laboratory for us to test various hypotheses and potentially answer some of the most important questions we can possibly ask. However, the Red Planet is not just the scientist's choice – it has been selected by the public as the best extraterrestrial world for humans to visit in the near future, too. When private organisation Mars One called for applications for future Mars-bound astronauts in 2013, over 200,000 people flocked to register their interest (with a still-notable 4,227 eventually submitting paid applications). Whether or not Mars One will actually make the journey, this flurry of interest is telling ... especially given the fact that the organisation is proposing a one-way trip.

There are a few key reasons why Mars is top of the list. One is pragmatism. Although a trip to Mars is long on a human scale, it's short on an astronomical one. It takes multiple years for us to even attempt to reach the outer planets, and that's just one-way. Our other near neighbours, Venus and Mercury, are far more alien and inhospitable worlds, and would be more difficult to visit and potentially colonise.

We've been observing the planet for ages – Mars has never been one to blend in. The planet confounded early astronomers with its movement through the sky and has contributed to some of the biggest shifts in scientific understanding in history. Mars is close to Earth and easy to locate in the sky; without a telescope, a keen-eyed observer is able to see five or six of the planets in the Solar System (Mercury, Venus, Mars, Jupiter and Saturn, and Uranus at a push). However, Mars's characteristics tell a particularly good story.

If you've ever looked up at the night sky when Mars is visible, you'll know just how red the pinprick of a planet really is. The planet's name stems from the Roman god of war, and it's easy to see why. The ruddy red of the planet's surface is reminiscent of blood, rust, hot metal, fierce flames and fire. It is a droplet of blood swimming in a dark cosmos. Throughout history, Mars has represented the traditional stereotype of masculinity, of anger, power, passion, strength and aggression. Various mythologies and civilisations have characterised Mars this way for century upon century, linking together the planet's astronomical properties with the human psyche, forming complex astrological and occult beliefs that still linger today.

With a simple telescope it's possible to pick out individual features on the planet's surface. Mars is a patchwork swirl of dark and light blotches ('albedo features'), stamped and peppered with craters, with little white caps at the poles. Using space telescopes such as Hubble we can even see signs of stormy weather if Mars is close and experiencing particularly dusty conditions. Mars's proximity to Earth and characteristic colour have helped the planet to cling on to its widespread appeal.

Venus is closer to Earth than Mars and appears brighter in the sky, but its surface is somewhat less interesting to the Earth-based observer; the entire planet is cloaked in a thick, toxic smog of gas that makes it appear smoother and featureless. Depending on the positions of the planets in space, on rare occasions Mercury can also be nearer to us than Mars, and the Moon is our ever-faithful nearest neighbour. However, we see these two bodies as dead, cratered, irradiated worlds, with Mercury being far too hot for even the most sun-seeking of organisms. Mars has far more potential.

For many years, we believed (and hoped) that Mars hosted developed life. After all, why should Earth be the only planet in the Solar System to have living beings scampering around on its soil? Early astronomers claimed to see a criss-crossing network of artificial canals built by thirsty Martians, who needed to funnel water from the polar ice caps to keep the rest of the planet alive. Many were so sure of this that the Pierre Guzman Prize, a 100,000-franc prize established in 1889 and awarded by the French Academy of Sciences to one recipient each in the fields of astronomy and medicine, was offered to the first scientist to find a way to communicate with another 'heavenly body' … other than Mars. Mars was considered simply too easy a target, and 'sufficiently well known'. The prize fund was eventually cleared by the *Apollo 11* astronauts in 1969 following their landing on the Moon – but if they'd landed on Mars, they wouldn't have received a cent.

Later observations showed bizarre shapes sculpted out of red rock, humanoid faces and structures akin to the ancient Egyptian pyramids, and even the potential ruins of a once-thriving city. Radio astronomers claimed to detect odd signals coming from the Red Planet. Could intelligent Martians be trying to signal to us?

Here, science fiction weighed in and furthered the excitement; authors excitedly penned (and later filmed) stories of a planet filled with aliens – little green men, aggressive robot warriors, thirsty canal-building architects, a civilisation of creepy human replicas, even an invisible collective consciousness or airborne hive mind. We've imagined ourselves both desperately trying to get to Mars and experiencing Martians coming to visit us (sometimes with catastrophic consequences). In a way, the word 'Martian' has become synonymous with 'alien', which would make it all the more significant if we were to find life on Mars.

This is perhaps the most pressing motivation to visit the Red Planet – the potential discovery of life. This is possibly the most important scientific question in existence. Are we alone in the Universe? Did life really only evolve on Earth? There are trillions of planets in our galaxy alone. Shouldn't one of these worlds also be home to some form of life?

Discovering signs of life on Mars would not just be the discovery of life on Mars, but the first discovery of any extraterrestrial life in the entire Universe. It would obviously be unimaginably exciting if life were thriving on Mars, unlikely as this may be (although it's still a real possibility). However, even if we were to 'only' find fossils or signs of past life, it would be truly significant. If two planets in the same Solar System had hosted life at any stage of development, it would indicate that the Universe may be full of life. Any planet (Earth-like or not) around any star in any galaxy might have its own set of little green astronomers scratching their heads over the very same questions humans have strived to answer for hundreds of years. We may even have already discovered a life-bearing planet around another star and observed it from afar.

'If life had originated twice, independently, within our solar system, we'd have to conclude that it can't be a rare fluke – and that the wider cosmos must teem with life, on zillions of planets orbiting other stars,' wrote Astronomer Royal Martin Rees in 2014 in an opinion piece for the *Telegraph*. 'But until we find life on Mars (or maybe on the moons of Jupiter or Saturn, or on a comet) it remains possible that life is very rare and special to our Earth. We now know for certain, of course, that there are no Martians of the kind familiar from science fiction – the kind that might provoke a "War of the Worlds". But the existence of even the most primitive life on Mars would in itself be hugely interesting.'

If we assume biological evolution to be a common and universal process, we'd expect microbial life to eventually turn into intelligent life of some kind over time, as we know happened on Earth. Finding evidence of life on Mars would therefore indicate that not only is life itself common, but intelligent life may be common, too. There's also the possibility of 'panspermia' – that life arose on some common body, say an asteroid or comet or another planet such as Mars, and then travelled to our planet at some point in history. It might have developed first on Mars and then been blasted out into space, sending Martian life hurtling to Earth on a meteoroid. If this version of panspermia were true, every single organism that has ever existed on Earth would technically not be terrestrial in origin, but Martian.

The signs are good that Mars may once have been a habitable world. It used to be far warmer and wetter than it is today. While there's a chance that primitive life might still exist, perhaps in the form of microbes locked up in subsurface water reserves or lying dormant in the soil, the chances of finding fossilised bacterial or microbial life are far higher.

Some scientists, such as astrophysicist Neil deGrasse Tyson, rate the odds of finding signs of past life at 50 per cent or higher. Others have said that, given what we've discovered about the planet's history, they're quite surprised that we haven't found any signs of past Martians yet. Yet others believe that life most probably did arise on Mars at some point and could still exist in pockets sitting deep underground. Either way, the Red Planet is the best possible site for our hunt for extraterrestrial life, especially if we'd like progress to happen in the near future.

Although the hunt for life dominates the rhetoric around Mars, it is far from the only scientific reason to go there. There's just so much science to be done on the Red Planet! Mars forms a quarter of our little terrestrial family, the only planetary family known to host life in the entire Universe – and the only family we have ever experienced. We believe that Mars once used to be very much like Earth, making it a sensible place to hunt for signs of our planet's history. It may be different in many respects, but Mars is still by far the most Earth-like planet in the Solar System. By exploring how Mars formed, and how it has evolved into the planet it is today, we can not only understand our neighbour better, but also learn more about ourselves – about how Earth formed, how rocky planets form and evolve in general, whether or not they are often habitable (if not inhabited) and, if inhabited, how such life came about. While Mars is geologically similar to Earth, it formed more quickly in the Solar System and has seen far less change across its surface, meaning that its geologic record is both much more complete and extends further back in time than our own.

Many prominent scientists and engineers believe that, all things considered, Mars is simply the best place to go to next.

These include Stephen Hawking ('Mars is the obvious next target'), Bill Nye, Elon Musk, Neil deGrasse Tyson, Carl Sagan ('The next place to wander to is Mars'), NASA administrator Charles Bolden ('Mars is a stepping stone to other solar systems'), and more. Former NASA *Apollo 11* astronaut Buzz Aldrin even created his own line of 'Get Your Ass to Mars' T-shirts, based on Arnold Schwarzenegger's famous line in the 1990 Mars-related film *Total Recall*. The aforementioned Elon Musk (former PayPal magnate, now of SpaceX and Tesla fame) has been outspoken about how humans need to explore and colonise other worlds, working towards the goal of becoming 'a multi-planet species'. This is another oft-cited reason for going to Mars: the survival of humankind. It may seem a far-off and abstract reason for such short-term urgency, but it's worth considering. A single global-scale event could wipe out our entire species. It could take us hundreds of years to build stable and functioning human colonies on any world in the Solar System, Mars included – so why not start now?

'Spreading out into space will have an even greater effect [than Columbus had when he discovered the New World],' said Stephen Hawking in a NASA lecture in 2008. 'It will completely change the future of the human race and maybe determine whether we have any future at all. It won't solve any of our immediate problems on planet Earth, but it will give us a new perspective on them, and cause us to look outwards rather than inwards. Hopefully it would unite us to face a common challenge. If the human race is to continue for another million years, we will have to boldly go where no one has gone before.'

Although the timespans involved are far beyond human comprehension – hundreds of millions to billions of years – there will come a time when our star dies and our planet is no longer the Earth we know and love. However, there is also a not-insignificant chance this could happen far sooner. Humans are a fragile species. Anything from a rogue asteroid to a volcanic mega-eruption, aggressive burst of cosmic radiation, biological pandemic, nuclear catastrophe or war, geological disaster, or drastically lower (dying out) or higher (gross overpopulation) birth rates could put an end to humankind. Something that we haven't yet invented may end up completely wiping out our species – rebellious artificial intelligence, perhaps, or a genetically modified pathogen. One of the most pressing predictable risks is that of climate change, of a runaway greenhouse effect that continues to warm our planet faster than we can adapt in order to survive.

'In this century, not only has science changed the world faster than ever, but [it has done so] in new and different ways,' said astronomer Martin Rees in a 2005 TED Talk. 'Targeted drugs, genetic modification, artificial intelligence, perhaps even implants into our brains, may change human beings themselves. And human beings, their physique and character, have not changed for thousands of years. That may change this century. It's new in our history. And the human impact on the global environment – greenhouse warming, mass extinctions and so forth – is unprecedented, too. This makes this coming century a challenge.'

Rees has given the human race a 50/50 chance of simply surviving the twenty-first century. Other scientists and philosophers have suggested somewhat more favourable odds, but none that are especially comforting – a 10 to 25 per cent chance of extinction within the twenty-first century, a 30 per cent chance within the next 500 years. This may sound like something from science fiction, but it could become science fact. It might be sensible to consider colonising another world if we want the human race to last, despite the immense cost involved. This is a key reason behind the push for humans to become a multi-planet species – pure survival.

'The Universe is probably littered with the one-planet graves of cultures which made the sensible economic decision that there's no good reason to go into space,' declared the ever-eloquent web comic xkcd in May 2011 (comic 893), 'each discovered, studied, and remembered by the ones who made the irrational decision.'

Another less terrifying reason is that of pushing ourselves, of driving progress for the entire human race, and of doing it predominantly for the challenge it poses. Whether you believe it to be worth the investment or not, space exploration is inherently inspirational. Walking on the Moon, orbiting our planet from above, setting off into the Solar System to explore worlds that have until now sat firmly in the realm of science fiction and speculation – it's exciting stuff. Many prominent scientists (and science enthusiasts) today attribute their passion for science and technology to the unrivalled space fever of the Apollo era in the 1950s and 1960s, and in turn have passed their joy for science and technology to subsequent generations.

However, this inspiration is not merely for the sake of it. Its effect is quantifiable. Science is interdisciplinary – the technologies, products and software developed and trialled when developing a space station, for example, can have an incredibly positive and profitable impact on the world. However, the question of money and financial worth is a very reasonable one, and possibly the strongest argument against going to Mars. Is it worth ploughing money into space exploration? Is 'for the sake of the challenge' a strong enough reason? After all, we have pretty substantial work to do here on our own planet. Wouldn't our money be better spent here at home?

Well, in a way, that's what we're doing. Our space programmes contribute a huge amount in terms of both money and advancement to our societies and economies here on Earth.

NASA's research is a great example of this; the different processes, programs and technologies that were perfected on the way to the Moon and in the decades since have had a far-ranging and significant impact back here on Earth. This research has either wholly developed or contributed to the development of artificial limbs, invisible braces, solar cells, water filters, cordless vacuum cleaners, mobile phone cameras, freeze-dried foods, firefighting equipment, scratch-resistant glasses, cancer treatment technologies, cochlear implants, thermal blankets, ear thermometers … the list is impressive. NASA research in the 1970s was also responsible for creating memory foam (or temper foam), the material used in unfathomably comfortable mattresses, which obviously makes the agency's entire budget a worthwhile spend! Some estimate that for every US$1 invested in NASA, the economy gets up to US$10 back (any returns from NASA's patents and licensed products belong to the government, not the organisation itself).

It's also worth considering that space science projects don't just comprise hunting for extraterrestrial life or studying how our planet formed billions of years ago, which are understandably of less importance to the average person. Continuing the focus on NASA, for example – NASA projects are diverse and include monitoring our planet's climate, burgeoning weather events and potential natural disasters, improving the quality of the US's aviation and air transport systems, and characterising space weather so we can avoid damage to our satellites (which themselves are used for everything from telephone to internet to GPS to military reconnaissance to scientific research).

Specifics aside, space agencies across the world are far from a drain on the economies of their home countries, instead acting as drivers for scientific and technological research, and inspiring the next generation of innovators. Funding a space agency necessarily funds scientists of all disciplines – engineers, chemists, physicists, doctors, geologists, computer scientists, biologists – all of whom don't work in space but here on Earth, contributing to societies and economies across the world.

'I really see [going to Mars] as an investment in our future, to inspire our young kids – and also, I think, to help our country and our economy for many years to come,' said former NASA astronaut Mike Massimino to Space.com in 2015. 'I think it would be a glorious thing to do.'

There are numerous reasons for going to Mars, and many believe them to outweigh the risk and cost involved. However, perhaps the strongest reason, in an emotional sense at least, is that of good old-fashioned curiosity. Curiosity, the drive to understand and explore both our world and the Universe in which we live (and the very apt name of NASA's fourth and largest rover), has driven many of the most monumental achievements in human history, from flying across oceans and continents just for the sake of it to understanding more about the nature of medicine or electricity or human biology.

One part of this meaning requires a global perspective – to believe that we are not just individuals and the centres of our own lives, but a member of the human race, the only known civilisation in the Universe and one that naturally longs to explore its surroundings (whether it be sailing to find the Americas, sailing to the Moon, or sailing to a neighbouring planet or beyond). As Stephen Hawking famously said in a 1988 interview with German magazine *Der Spiegel*: 'We are just an advanced breed of monkeys on a minor planet of a very average star. But we can understand the Universe. That makes us something very special.'

Space science tackles the biggest questions we can think up: are we alone in the Universe? How, when and where has life formed in the cosmos? Is it unique to Earth? If there is life, what does it look like, could we communicate with it and where is it? How did our planet and Solar System form? Could we possibly travel to, and colonise, another planet? How would we do it, and what challenges would we encounter? Could we overcome them?

Mars is one of the only destinations we are currently able to explore in order to answer some of these questions – and more.

CHAPTER TWO

The Wolf and the Woodpecker

Space is a beautiful place.

It's hardly surprising that ancient astronomers judged the cosmos by appearances. They used their early preconceptions of the Solar System to characterise their world and draw meaning from the heavens. This is especially true when it comes to Mars. The planet's reputation throughout history has been based on pretty much one thing: its redness.

To fully understand the basis of our modern-day opinions about Mars, it's worth briefly delving into various aspects of the colour itself. Most, if not all, of the planet's modern-day connotations stem from its colour, but are also linked to – or have influenced – Mars's various appearances as a deity and mythological figure throughout history.

Red is an important colour in human culture and psychology. Today, we judge the colour red as being brash and in-your-face. It's often used in marketing techniques to promote action; web banners and online adverts often flash red to attract attention and draw your eye, enticing you to click through and spend your hard-earned cash. In other ways, the colour represents anger, rage, fury, wrath, energy; one 'sees red' when they're furious or enraged. In some languages (Hebrew, Inuit, Sanskrit), the word for 'red' is – or once was – interlinked with that for 'blood' or 'blood-like', reinforcing its connection to violence and aggression. Red has long been associated with evil or danger, and used as a sign of foreboding and warning (often literally, in the form of stop and danger signs, and via symbolism related to the devil). It's also commonly linked to sex and passion, from red-light districts to the sickly hue of Valentine's Day merchandise, and the societal sex appeal of crimson lipstick and red dresses (a real and intriguing phenomenon known as 'the red dress effect'). On a lighter note, it's also the colour of luck, celebration and joyful anticipation, especially in Asian countries (notable examples include Chinese red envelopes and red egg ceremonies for new babies).

Some of this is based on social conditioning, some on scientific principles and some perhaps on a mix of both. For example, flashes of red in nature stand out very clearly from most habitats and backgrounds (leafy, desert, snowy, aquatic, even unnatural cityscapes) and red light is the least scattered by molecules in the air, allowing it to travel further and be more easily visible at greater distances. As such, it is ideal for signalling danger or warning. It's used in this way by many animals to deter predators (something known as 'aposematism', where an organism such as a frog, snake, insect or mollusc uses a physical characteristic – colour, smell, noise – to stop itself from being eaten; the most commonly used colours are red, yellow and black). Red is also often naturally linked to sex, as with a baboon's scarlet posterior or the rosy flush of aroused skin. High testosterone levels can tint skin and faces slightly red or ruddy in some species – birds, fish, primates – thus linking the colour to masculinity (given that male levels are higher than female ones). In the same vein, poor health leaves skin pale and washed out, while pinky toned skin seems to glow with health.

We react to colour significantly and subconsciously. Various studies into colour psychology in western marketing have shown that consumers link red to power, excitement, lust and love, and that this actually influences our purchasing habits. Both sexes are more attracted to red, and perceive those wearing or even being near the colour (for example, in a red room or car) as being more attractive and interested in sex. Researchers have even found that sports teams or competitors who wear red kits are consistently more likely to win than their non-red counterparts – and gamers assigned to red teams appear to be more successful in the same way. The reasoning for this is thought to be psychological; red signals dominance and strength, which may subconsciously affect the competitors both positively and negatively (winning team is confident, losing team is intimidated). Next time you're placing a bet, it may be worth rooting for the red team.

Numerous ancient civilisations have associated red with the aforementioned qualities in mythology, folklore and culture. The Ancient Greeks, for example, linked red – and, by extension, Mars – to traits that are still considered traditionally masculine. It's thought that red was also associated with their army, making it the colour of war, bloodshed and aggression. A number of older civilisations greatly valued their armies, relying heavily on their reputation as successful warriors to survive and thrive, and thus held Mars – as both a deity and a planet – in particularly high esteem. Homer's *Iliad* is a particularly famous body of ancient work that aligns with this Ancient Greek perspective; in the *Iliad*, Homer describes Mars as 'monstrous', 'insatiate of battle' and 'murderous'. 'Fighting can delight the heart of Mars'!

However, red is not solely associated with traditionally masculine elements; in Chinese culture, of the five essential elements of being (metal, wood, water, fire and earth) red represents fire, adding to its vigour, energy and impulsiveness, but the colour is predominantly linked to happiness and good fortune.

More generally, ruddier, rustier hues of red have long been used to convey a sense of fertility and virility, and were associated with nature and the season of autumn (harvest). The astrological connection between red, Mars and fertility is strong enough that the springtime month of March, which often sees the emergence of new life (in the northern hemisphere at least), is named after the planet.

This is especially true in Japan, where red links to life and love, India, where it represents love, seduction, beauty and purity, and in ancient Egypt, where red dyes were often used in both beauty products and cosmetics, and to celebrate vitality and health. This natural association may have its roots in the practicality of the colour red; various civilisations and societies across the world used different forms of dye to manufacture red clothing, cosmetics and dyed products, all of which were natural in origin (iron-rich clays, hematite minerals, cinnabar, safflower, henna, insect shells). The colour red seemed to spring from the very earth – it was provided by Mother Nature.

Perhaps surprisingly, one modern culture to steer clear of pure aggression is Russia, for whom red represents beauty, joy and vibrancy – the Russian word for red (*krasny*) is closely linked to that for beautiful (*krasivaya*). However, there are traditional elements of red in Soviet and Russian culture; red connotes blood, life, birth and high status, and is also famously linked to communism via the Red Army, who adopted red as their colour to represent the blood of their members. (While red still most often represents left-wing ideologies, it's interesting to note that it is also used today by the American right-wing Republican party, muddying the association.)

This is not just a fanciful stroll through colour preferences and abstract symbolism. Every single aforementioned attribute of the colour red has also been astrologically associated with the planet Mars, purely due to its obvious redness.

This redness is actually real and is caused by the composition of the rocks sitting on Mars's surface. The planet's surface rocks are iron-rich, causing them to slowly oxidise – in other words, rust – and turn the same orange-red hue as rusty metal does here on Earth. Dust and small powdery particles stream off from these rocks and permeate the planet's atmosphere, tinting it a salmon or pinky red colour that is observable from afar. In reality, Mars's surface is quite diverse in colour, with vast regions tinted in tan and yellow 'butterscotch' hues through to reds, oranges, browns and even olive-green tones – but that's not what we see from Earth.

To stargazing civilisations throughout history, the Red Planet's rusty colouring resembled the superheated metal in a blacksmith's forge, the hue of spilt blood on a battlefield, blazing flames and fire, the violent haze of anger and rage. It's no surprise, then, that in the astrology and mythology of most cultures the planet Mars primarily represents dominance, aggression, anger, passion, impulsivity, sexuality, power, ambition, action, strength, virility and fieriness – quite a single-minded list! In other words, Mars is the very embodiment of the stereotype of traditional masculinity (which, while tired and outdated, remains annoyingly persistent).

The Romans named the planet after their god of war and weaponry, Mars (Mars was also an agricultural guardian, linking back to virility and fertility). The Roman Mars is the counterpart to Greek mythology's equally destructive and virile god of war, Ares. Although they stem from different societies and cultures, both Greek and Roman mythologies are similarly structured and have roughly interchangeable characters (as with Mars and Ares). Some of the more famous Greek–Roman counterpart pairings are Zeus and Jupiter (kings of the gods), Hera and Juno (queens of the gods), Poseidon and Neptune (gods of the sea), Hades and Pluto (gods of the underworld), Eros and Cupid (gods of love) and Aphrodite and Venus (goddesses of love).

In Roman mythology, Mars is usually depicted as a muscular warrior with a cloak and plumed helmet, sometimes nude or partially nude, holding a shield and spear. In Roman art he is usually older, bearded and wearing battle armour. In Greek-inspired depictions, he is young, clean-shaven and often nude or scantily clad (a bit rash given the sharp weaponry he carries, but apparently symbolising his fearlessness in the face of danger). Think Gerard Butler in the film *300*, whose wardrobe essentially blended together various mythological depictions of Mars and Mars-like characters (though admittedly with quite a few inaccuracies and tweaks, apparently done intentionally for the sake of appearances and ease of filming). Mars's shield and spear also give rise to his symbol, which doubles as the symbol for male – a circle with an arrow pointing to the upper right.

In Greek mythology, Mars/Ares had twin sons with the goddess of love, Aphrodite, named Phobos and Deimos*. Appropriately, these names were given to the planet's two

---

* Although they are most commonly described as being the sons of Mars, some sources – including the International Astronomy Union (IAU), the body responsible for naming cosmic objects – state that Phobos and Deimos were named after the two mythological horses that pulled Ares/Mars's chariot. Either way, they are referred to as the 'attendants' of Mars and drove his chariot in one way or another.

small moons when they were discovered in the late 1800s. Phobos, the Greek word for phobia, represents fear and panic, while Deimos, the Greek word for terror, is the personification of dread and flight (especially related to war, for example retreat on the battlefield, or fleeing from defeat). Overall, the pair depict terror, fear and loss – unsurprising given the identity and aggressive nature of their warlord father!

As well as emotional meaning, various mythologies also associated specific animals with their gods. Many choices are obvious – hunting animals and predators (wolves, birds of prey, bears), stubborn and fiery animals (goats, rams), animals perceived as symbols of power and status (horses) and venomous or deadly creatures (snakes, spiders, scorpions).

The two animals the Romans linked to Ares/Mars, for example, are the wolf and the woodpecker. This may seem like a bizarre duo, and the latter a particularly left-field choice, but the early inhabitants of Italy held the woodpecker in high esteem. Despite what Woody may have taught us as children, the woodpecker is strong and intelligent. The species thus held a high status in the world of birds and was especially important in the ancient art of augury, in which the flight and behaviour of birds was studied in order to understand the will of the gods – a kind of divination taken very seriously by ancient societies. Woodpeckers were also thought to play a role in the fertilisation of springtime crop fields, linking to Mars's reputation as an agricultural protector, virile and full of 'life force'.

However, there is another side to this coin: despite their high status, woodpeckers were apparently a common food source in Ancient Greek society. Greek historian Plutarch (46–120 AD) wrote about this in his collection of essays on morals and ethics (entitled 'Moralia'). He questioned why some were so averse to eating woodpecker when it was such a normal thing to do at the time, also wondering why, exactly, the woodpecker was chosen to represent the planet Mars. 'Why do the Latins revere the woodpecker and all strictly abstain from it – is it because they regard this bird as sacred to Mars?'

One explanation of the high status of the woodpecker, suggested Plutarch, is the tale of Picus. Picus was an important figure in Roman mythology. He was said to be the first king of the Italian region housing modern-day Rome, the god of agriculture and farming, and a Roman son of Mars (though he is sometimes considered to be a son of Saturn). After Picus scorned the charms and advances of a witch, she transformed Picus into a woodpecker in a fit of rage. (As a nice ornithological aside, woodpeckers are of the genus *picus* – for example, *picus viridis* is the Latin name for the European green woodpecker.)

Plutarch's second explanation is the tale of Romulus and Remus, which also explains the unusual woodpecker–wolf pairing. This is one of the most famous Roman myths and tells the tale of two feral boys raised by wolves. The boys, said to be the aforementioned twin sons of Mars, are abandoned in a river to die shortly after their birth, since their rightful claim to power posed a threat to the leader at the time (Amulius, their mother's uncle). However, the river sweeps them to safety and they are discovered by a she-wolf, who suckles them to keep them alive. The twins are also regularly visited by a woodpecker that brings them scraps of food. After a while they are found by a shepherd and his wife, raised to adulthood and become key figures in the founding of Rome (which is named after Romulus himself). In some accounts of the fable, the aforementioned practice of augury (bird divination) is relevant in the founding of Rome; each twin preferred to build the city on a different site and agreed to use augury to choose between them. However, the two quarrelled over the results and Remus was killed, leaving Romulus as the sole founder – and namesake – of Rome.

What's in a name?

Although dominant in the English-speaking world, Mars is far from the only moniker for the Red Planet. Ancient Chinese mythology paints Mars as a fiery, energetic, choleric influence based on its association with their 'element' of fire.

The old traditional Chinese name for Mars was 熒惑星 (*yínghuòxīng* in pinyin), which has a meaning along the lines of luminous, shimmering, firefly, lightning bug, glow-worm. *Yinghuo* was used as the name for the first intended (and sadly unsuccessful) Chinese mission to Mars in 2011, *Yinghuo-1*. The modern simplified name for Mars in China is 火星 (*huǒxīng*), which bears approximately the same meaning – these characters are also used in Japan, which refers to Mars as 火星 (*kasei*, literally 'fire star'), Korea (火星, *hwaseong*, or 화성 in the modern Korean language of hangul), and other parts of Asia*.

In Hebrew, Mars is named *Ma'adim*, literally 'turning red' or 'blushing', while in Hindu astrology it takes the name *Mangala* (also *Angaraka*), after their flame-red god of war (whose name literally means something akin to 'one who is red', or 'the red one'). In various forms across the world, these words for Mars have given rise to the modern-day names for the second day of the working week – in English, the word 'Tuesday' derives from the name of an old Norse god named Tiw or Týr, the Germanic equivalent of Mars as a god of war and figure of aggressive victory. This pattern – of Tuesday deriving from a Mars-like figure – has continued in various cultures. In Japanese, for example, Tuesday is 火曜日 (*kayōbi* in *romaji*) and in Korean 화요일 (*hwa-yo-il*), both 'fire day'.

The Babylonians associated Mars with Nergal, their god of war, fire, destruction and the underworld. The ancient Egyptians – who maintained some of the earliest known astronomical records – named the planet *Her Desher* (or *Har Dasher*, 'the red one') or Horus the Red, in reference to one of their most important and esteemed deities (Horus, who had the head of a falcon and was thus connected to the skies). There are some nice parallels between the different cultures

---

* Chinese characters are also used in the Japanese *kanji* and old Korean *hanja* writing systems, but pronounced and combined differently. Thus '火星' refers to Mars in Chinese, Japanese and (traditional) Korean, but is said as *huǒxīng*, *kasei* and *hwaseong*, respectively.

here. For example, Mangala rides on a ram, which in turn represents Aries in the celestial system of the zodiac. Aries is ruled by the planet Mars and thus, predictably, associated with the element of fire. It's easy to see where the traditional astrological description of an 'Aries personality' – impulsive, brash, courageous, passionate – comes from!

The ancient Egyptians also referred to Mars moving backwards in the sky via the moniker *sekhed-et-em-khet-ket* ('he who travels/moves backwards' – rolls off the tongue!), something that may be responsible for the idea of Mars being impulsive, rebellious and non-conformist. This backwards motion, known as 'retrograde', has been observed for about as long as humanity has been gazing at the skies and was dutifully noted in various ancient astronomical records.

Retrograde motion is simple in theory, but can be difficult to visualise. It all boils down to the motions of Earth and Mars relative to one another. We see the sky – Sun, stars, planets and all – to rise in the east and set in the west. This isn't a motion inherently applicable to the heavens, but a product of Earth's direction of rotation (anti-clockwise, if we were to look down on it from the North Pole – only Venus and Uranus rotate the other way, a phenomenon known as 'retrograde rotation'). If we were to rotate clockwise, we'd see the opposite apparent motion in the sky. Likewise, we see Mars to move across the sky from east to west – most of the time.

Roughly every two years Mars performs a giant loop-the-loop, quickly swapping direction and scooting westwards. For a couple of months it continues along its strange path, before abruptly halting and resuming its normal eastward motion. Rather than just being an indecisive rogue planet, bucking the trend and confounding astronomers for fun, this is an optical illusion based on the motion of two planets relative to one another.

Mars takes just under 687 days to orbit the Sun, while Earth takes just over 365. Picture the two planets on an oval Scalextric track, with Earth nabbing the inside lane. In the amount of time it takes Mars to loop the track once, Earth is coming close to completing two circuits – meaning that at

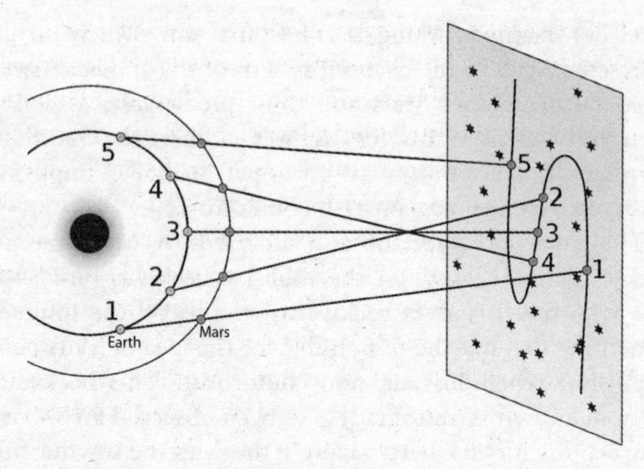

Diagram explaining the phenomenon of retrograde motion – the real motion of the Earth (inner ring) and Mars (outer ring) is shown on the left, with Mars's resulting apparent motion seen in our sky on the right.

some point in orbit, Earth has overtaken its red rival. The moment at which it overtakes is when the apparent retrograde motion takes place. As Earth catches up with Mars, the planet seems to briefly halt. As Earth moves past Mars, briefly overtaking it on its elliptical path around the Sun, Mars seems to fly backwards in the sky – in reality, it is doing no such thing. As Earth carries on around its loop, Mars 'catches up' to our perspective and appears to halt briefly before carrying on 'eastward' as before. This retrograde loop usually takes Mars a couple of months to complete.

If we were on the exact same orbital plane as Mars, we would see the planet move backwards and forwards along the same line of motion, like drawing over a single line twice with a pen. However, as Mars's orbit is slightly tilted with respect to our planet, we see a different scene play out depending on how the two planets are orientated at that point in their orbits: sometimes it's a loop, sometimes a zigzag, sometimes a curling 'S' shape. The 2003 retrograde motion, for example, zipped by as a loop-the-loop, whereas the 2005 event resembled a more jagged 'Z'.

The concept of retrograde motion is a good example of how the practices of astronomy and astrology are strongly intertwined and have been for centuries. Retrograde motion is a solidly scientific astronomical term. However, the concept is also used in an astrological sense. Astrologers frequently refer to various planets (often Mercury) as being 'in retrograde'; rather than referring to the aforementioned cosmic trick of perspective, this phrase has a distinct astrological meaning – that of reversing your characteristics, challenging your stereotypes and becoming more introspective, reassessing your life goals and desires, of things becoming muddled or unclear, or seeming to regress or slow down. It's relatively evident where these claims stem from!

Astrology plays a starring role in humanity's relationship with Mars. Astronomy, or the scientific study of the Universe, is a truly ancient science. It has been practised the world over by pretty much every civilisation for many thousands of years. However, the practice of astrology isn't much different; despite the frustration it can cause for astronomers, there is a reasonable explanation for why astrology and astronomy are often confused. Although astrology is now firmly placed in the pseudoscience category, this is a relatively modern shift that only happened a few hundred years ago.

Astrology and astronomy were closely interwoven for many centuries, with astronomy thought to be the theoretical science underpinning the practical application of astrology. Many eminent astronomers – Galileo, Copernicus, Kepler, Brahe, Cassini – were also highly respected astrologers of their day. In fact, some countries still offer advanced degrees in astrology from well-reputed academies and institutions, India being a notable example (some suggest that this is due to the lack of widespread education in such countries, which allows astrology to still be counted as legitimate among the sciences).

In a nutshell, astrology is based on the idea that the human world is influenced by the celestial one. There are numerous different strains of astrology, dealing with everything from the weather through to health, agriculture, human personalities and more. While some of these make some sense (weather, for example, when you consider that the Sun and Moon do affect some of the phenomena we experience on Earth), other areas are more questionable (most prominently horoscopes and the concept that the alignment of the planets at the time of our birth can affect our moods, characteristics, personalities and everyday lives many decades later).

The seventeenth and eighteenth centuries saw a time of great discovery and change in Europe, with physics and astronomy in particular progressing at an incredible rate. Astrology slowly lost support throughout the seventeenth century and had been mostly abandoned in a scientific sense by the eighteenth. However, it was still loved by followers of occultism – spiritual practices such as magic, palm-reading, tarot cards, clairvoyancy, alchemy, mediumship, some religions, divination and more – and still is today.

Rather than dismissing it outright, scientists have made genuine forays into astrology to test the tradition's veracity and legitimacy. One attempt to preserve its value – in a largely pragmatic sense rather than a scientific one – was Jung's development of 'psychological astrology'. Jung saw astrology as a kind of ancient form of psychology, claiming that all we had ever learnt about the human psyche and condition, self-awareness, mentality, thought and so on had been well-documented within astrology. 'Its value is obvious enough to the psychologist, since astrology represents the sum of all the psychological knowledge of antiquity,' said Jung in 1930.

Later came studies into the phenomenon of the 'Mars effect', spearheaded by French psychologist Michel Gauquelin in the 1950s. Gauquelin claimed that Mars was in the same position in the natal charts for a significant number of exceptional athletes – in other words, a statistically high proportion of athletes were born shortly after Mars rose in the sky. For many years other scientists probed this apparent correlation, with results coming out both in favour of Gauquelin's hypothesis and in strong disagreement. Unfortunately, it turns out that we can't track Mars's motion through the sky in order to produce a super-athletic brood of children. In 1997, a team of researchers concluded, 'after persistent and painstaking examination', that 'there is insufficient evidence for the "Mars effect" ... this effect may be attributed to Gauquelin's selective bias in either discarding or adding data *post hoc.* It is time to move on to other more productive topics'.

The 1970s saw the arrival of the New Age movement, and with it the now-unavoidable flood of tabloid horoscope columns and television astrologers. These columns, while (hopefully) very rarely regarded as rigorous, draw on our established astrological perceptions of Mars. It's likely that your first opinion of Mars relates to masculinity or war and, in general, we do still perceive Mars as a stereotypically aggressive concept – although perhaps not quite as explicitly so as we once did. Just as Mars is a very real lump of rock whirling through space, there is also the concept of 'the astrological Mars', which has a distinct definition and meaning on the astrology scene.

'Although we do consider the qualities of the actual planet, astrologers are much more interested in the symbolic,' explains John Marchesella, president of the National Council for Geocosmic Research (a US organisation dedicated to improving astrology education and research worldwide). 'We're more likely to turn all the physical qualities into symbols and, certainly, all the mythological meanings are already symbolic. Mars is covered with iron dust, for example, and in mythology, he ruled over the Iron Age, the age of weapons and tools. Mars is the "red" planet and red is often associated with war.'

The astrological Mars represents the individual's motivation and ability to fight for what they want, says Marchesella. The planet's defining characteristics include 'anger, ire, directness, forthrightness, the ability to defend one's boundaries, limits, territory, possessions and values ... Mars is our degree of speed, impulsiveness, hot-headedness. It symbolises competition, our degree of bravery, courage, valour. When Mars interacts with other planets, he speeds up or excites their nature.'

So far, so aggressive, as expected – but it has not always been quite so clear-cut. 'In historical astrology, Mars was seen not as positive or negative, but rather as a "malefic" influence,' adds Marchesella. 'Being the god of weapons, Mars had the ability to "cut" away from the clan or to cut the actual clan itself, neither of which was seen as a positive by ancient cultures – the name of the game was safety in numbers.'

It seems unfortunate that Mars got lumped with mostly aggressive characteristics purely because of its colour. Although astrologers claim that Mars's energy and vigour can apparently be channelled in myriad positive ways, being a hot-headed aggressor is not generally a good thing in modern society. However, astrology is certainly not discriminatory; the other planets also have meanings based mostly on a single aspect of their behaviour or appearance.

Mercury is defined by its incredibly speedy 88-day orbit around the Sun; it governs things that seem to move or shift – communication, travel, thought, adaptability. Venus is similar to Earth in several ways, leading to its reputation as our 'twin' or 'sister' planet. This bleeds into its astrological meaning – Venus governs empathy, balance, harmony, alongside love and beauty. Venus is the only planet in our solar system named after a female figure and its symbol is also the sign for 'female' (a circle with a cross beneath it). Jupiter's bulk casts it as a protective force, linked to prosperity and good fortune. This may have originated in astronomers' belief that Jupiter's gravity acts as a cosmic shield for Earth and the inner planets, capturing and slinging away potentially devastating comets and asteroids. Saturn is, predictably, defined by its rings, which apparently represent human limitations and boundaries, social structure, authority and hierarchy. Uranus rotates on its side, bucking the trend followed by all of the Solar System's other residents – and so of course it links to individuality, progressive and radical ideas, genius and revolution. Neptune and Pluto, discovered later than their siblings, seem to have more of a connection to the social or political movements taking place at the times of their discovery. Neptune represents the creative and compassionate, but also the confusing and deceptive. Its distance signals a separation or retreat from society – places like hospitals, monasteries, even prisons. Discovered in 1846, Neptune links to independence and utopia, likely influenced by the growth of different belief systems in the eighteenth and nineteenth centuries such as nationalism, socialism, communism and the welfare state. Technically speaking Pluto should be omitted from this section, but its story is interesting and slightly ironic – Pluto represents transformation, or digging beneath the surface to reveal the truth, regardless of how disruptive that truth may be. A self-fulfilling prophecy!

Astrology may be the poster-child of occultism, but the same symbolism used in astrology also crops up in many other places – in witchcraft, tarot and palmistry, for example, three other well-known spiritual and occult practices.

Just like astrology, witchcraft and magick are staunchly pseudoscientific ('magic' apparently refers to sleight-of-hand and stage tricks, while 'magick' is the real deal – the more you know). However, the practices' roots go back to when many modern societies were just beginning to form (notably the ancient Egyptians, Babylonians, and ancient Near East) and are recorded by many religions (notable examples include the Christian Old and New Testaments, which heavily condemn 'sorcery', and Judaism and Islam). Support for this kind of divination still exists today, although on a far smaller scale.

Contemporary witchcraft follows the usual pattern of astrology, associating Mars with vigour, energy, willpower and victory, also adding various herbs and crystals into the mix. According to the amazingly named online 'witchipedia', 'thorny and/or red plants are often associated with Mars, as are those with a strong, spicy flavour and the ability to warm and stimulate or energise the body. Any red stone can be used to represent Mars, including ruby and garnet. Bloodstone [a mineral otherwise known as heliotrope, which is flecked with red chunks of iron oxide] also contains

Mars energy'. In other words, we're back to Mars equals red, disruptive and energetic. Any plants or foods that are red, spicy, stimulating or related to blood are thus given to Mars – chilli peppers, raspberries, radishes, cranberries, garlic (which 'removes toxins and cleanses your blood' – pure pseudoscience, so please don't rely on this as diet advice), root ginger, bitter coffee, black pepper, paprika, any kind of curry spice, whisky, so-called aphrodisiac foods (again, not diet advice – 'aphrodisiac foods' largely rely on the placebo effect) and so on.

To connect to your 'Mars energy', a magick practitioner should integrate the colour red into their clothing and surroundings, perform Mars-specific rituals on a Tuesday (Mars day!) and light candles to bring fire into the room. If you fancy summoning Mars energy into your life, try wrapping yourself in red when you have a free Tuesday evening and enjoy a feast composed of spicy or red foods (bonus marks for both). Then, carve the glyph for Mars – a circle with an arrow pointing to the upper right – into a red candle using a pin. Light it and any appropriate incenses you might have (essentially anything pungent, bitter, red, or with 'blood' or 'heart' in the name), and meditate to tap into your inner sense of victory. Even if your magick isn't as effective as you'd like, it's a good excuse to enjoy a strong coffee, hot curry and dram of whisky, so you've nothing to lose!

Palmistry – or chiromancy – has its roots in ancient Hindu astrology and is widely practised in much of the eastern half of the world. The palmist map of the human hand charmingly makes it sound a lot like Martian terrain. Looking at your hand palm-up, the region just above the thumb, where there's a bit of fleshy skin, is called 'Mars Positive'. The corresponding region on the other side of the hand is called 'Mars Negative'. A small line curving downwards around the thumb – not the most prominent line, but a smaller and shorter one just within it – is the 'Line of Mars', and the patch of skin at the very centre of the palm is known as the 'Plain of Mars'. Mars Negative, Positive and the Plain come together to form the 'Mars Galaxy'.

The various Mars regions on the hand help one to analyse their temperament, levels of self-control, confidence, temper and drive to succeed – all qualities linked to the idea of strength, aggression and courage, a running theme for the Red Planet. While Mars can signal many things in palmistry, the overarching theme is more of the same: aggression, courage and fiery anger. 'If there are certain marks on the area of Mars, which is just above the thumb, it can make the person a lot more courageous, and they would often have some experience in the army or involvement with the military,' says London-based palmist Julia Dashkovskaya. 'Or, depending on what the rest of the palm is like, it can make them capable of aggression, anger and inflicting pain. The line of Mars can give extra fighting spirit in life if it's present.'

The other planets also play a large role in palm-reading. The fleshy 'mounts' underneath your fingers – they can be thought of as the undersides of your knuckles – are associated with Jupiter, Saturn, the Sun and Mercury respectively, from index to little finger. The prominence of these mounts can apparently tell us about promiscuity, leadership capabilities, patience, creativity, intellect and more.

Mars also features in tarot. In many individual tarot readings the planets aren't explicitly used, but the idea of Mars is linked to specific cards in a tarot pack, says Cilla Conway, an author, artist and tarot consultant. 'All cards have positive and negative aspects, depending on how the cards fall and what surrounds them.'

Traditional tarot decks come in packs of 78. A pack has four suits that are similar to a traditional pack of cards; each suit contains the standard cards, but with an added face card known as the Knight, which slots between the Jack and Queen. On top of this is another 21-card trump suit (a suit that becomes more powerful in certain scenarios) and an additional card called 'the Fool', which is the tarot equivalent of a lone Joker. Although they are now mostly associated with spiritual divination, these decks were originally used to play Bridge-like trick-taking games (most obviously, a French card game named Tarot). Together, the 21-card trump suit and the Fool make up the subset of cards used for tarot divination and occultism (called the 'Major Arcana'). These cards have weird and wonderful illustrations daubed on them – although tarot decks have been around since the fifteenth century, the tarot illustrations that are popular today date back to a deck released in 1910. This deck is bold and bright, full of yellows, oranges and reds, with characters including the Sun, World, Star, Lovers, Emperor and Death.

One of the deck's more menacing cards shows a dark, sinister tower set against a rain of lightning bolts and dark clouds. Two people, contorted in agony, are falling from the top of it, tumbling away from jagged flames. This card is called the Tower, and – surprise, surprise – is associated with Mars. The second card traditionally linked to the planet is Strength, or Passion, which shows a robed figure caressing the jaws of a lion. Together, these cards represent the traditional Martian influence – aggression, destruction, animal lust, strength, courage and control.

Mars also rears its head in the form of a card named the Emperor, a figure filled with masculine energy, representing action and pure courage. The Emperor card depicts a scary-looking man with a fiery red robe, golden crown and flowing white beard, sitting authoritatively on an imposing throne that has rams' skulls carved into the armrests and back. He holds a rod in his right hand shaped like an Egyptian ankh and a golden orb in the other. The rams hark back to the traditional Mars-dominated zodiac sign of Aries – another nice link between tarot and astrology. 'Mars certainly underpins the meaning and interpretation of the Emperor,' explains Conway. 'In traditional terms the Emperor represents the divine masculine, certain of his power and point of view.'

Regardless of belief system, Mars appears to represent the same emotions, characteristics and meanings wherever you look. War was a key part of ancient societies. This placed the planet in a particularly powerful and meaningful position, and it has continued to weave a great narrative ever since. Whether or not astrology, tarot and general occultism is your bag, the ancient occultist arts and practices do reflect our long-held opinions on the planet as both a cultural figure and source of symbolism. After all, we've only had the technological ability to look at Mars in detail via telescope for a few hundred years, and we've only had the capability to visit the planet for a tiny fraction of that time (about half a century or so). Before this, our thoughts about Mars were based on ancient beliefs and symbolism, which were in turn based on what little we knew about the planet's physical characteristics (predominantly its red colour and retrograde motion).

It's intriguing to see how our increasing knowledge about Mars feeds into the beliefs of the occult. Some witches draw a link between Mars's physical properties and its magical ones – for example, its thin atmosphere (Mars doesn't trap personal energy and doesn't hold grudges), its cratered and windswept surface (it can change and adapt), its colour (its 'element' is fire, representing passion and intensity), its iron content (a substance known as 'Water of Mars' or 'War Water' is made by allowing nails to rust in water and used to fend off aggressors) and the frozen water seen at its poles (it suspends emotion and doesn't get bogged down).

As a disclaimer, astrology is pure pseudoscience, as are the practices of the occult – while it may be somewhat fun to try to connect with your 'Mars energy', none of this should be taken as scientific or legitimate on any level. However, while we cannot learn about Mars as a planet from exploring its roots in myth and legend, it is undeniably interesting, especially in the context of modern society.

Mars's relentless masculine aggression makes sense when considering an ancient group of warriors or a society that must be battle-ready to survive (as Marchesella mentioned, when preventing your clan from being cut or torn apart was a real and everyday concern), but there's little need for such behaviour in many occult-practising countries today, where the traditional stereotype of masculinity is also being challenged more and more with each passing generation. However, some are happier about this than others.

'The pendulum appears to have swung too far away from Mars now,' Conway laments. 'Too many young men nowadays seem to have lost the potency of Mars and become soft and squishy.'

CHAPTER THREE

Marvin and the Spiders

Setting foot on the surface of another world. Gazing back at the Earth, a tiny 'pale blue dot', through the inky blackness of space. Reaching out into the Solar System to explore our nearest cosmic neighbours. Hunting for – and maybe even discovering! – extraterrestrial life.

These were the goals buzzing through minds across the world in the lead up to the Space Race. In the 1950s and 1960s, the world was gripped by space fever. This wave of obsessive, contagious excitement about space travel was almost palpable. This period was the Apollo era – the age of Apollo! The incredible technological progress made during this period ultimately culminated in humans successfully setting foot on the Moon, and has been invaluable in developing and advancing a whole host of other scientific fields since.

We're now experiencing something similar with Mars. Seemingly every month a space organisation or agency releases new or updated plans for a mission to Mars, and numerous space probes are currently scheduled or slated to begin their journey to the planet in the next couple of decades. While the Mars generation – today's teenagers and young adults, who missed all the Apollo excitement and can reasonably expect to see humans land on, and perhaps colonise, Mars within their lifetimes – is relatively young, the age of Mars has been around for longer than you might think.

Mars's rise to pop culture power over the other Solar System planets began decades ago. Even for those lacking a deep-seated interest in science, Mars is inescapable and has been for many years. Film, television, comic books, radio, games, music, art – seemingly every facet of human culture has some sort of connection to the Red Planet, or has been influenced by it in some way.

One of the first modern societies to become preoccupied with Mars was the Soviet Union. At the start of the twentieth century, the USSR was beginning to develop a near-obsession with spaceflight. Some members of society began weaving together a few threads of thought into one overarching philosophy – namely that of 'cosmism' (the idea that humans should travel to and colonise the cosmos, and that humans had an intrinsic relationship with the Universe) and that of 'technological utopianism' (the belief that rapid industrialisation and scientific advancement would ultimately create a utopian society).

This sparked a countrywide lust for interplanetary flight, a drive to visit the stars, a strong public interest in space and manned spaceflight – in a way, a sort of space activism movement. At the time, this was arguably based more on spiritualism and philosophy than it was on actual scientific or technological ability, and faded over the course of a few decades, but it had a few long-lasting and significant consequences. Most importantly, it moved the idea of spaceflight from the realm of fantasy into that of industry, likening it to an achievable goal like aviation and removing a level of mysticism. Spaceflight was no longer a fantastical endeavour but rather a logical, and completely feasible, step for an advancing and developing society. To build a strong and forward-thinking society, one must conquer the cosmos.

While the fervour of 1920s space activism inevitably began to fade, the belief that spaceflight was vital in building a successful society lasted for a long time. Perhaps as a consequence of the views of the early 1900s, the Soviet Union was one of the first countries to attempt spaceflight; from the 1930s onwards, they developed a series of test programmes and development projects aimed at launching humans to the stars. While we often think of the US as having simply 'won' the Space Race by successfully landing on the Moon first (1969), the USSR achieved an incredible number of successes along the way.

Among myriad other milestones, the USSR launched the first ever satellite (*Sputnik 1*, 1957), animal (a mongrel named

Laika in 1957) and human (Yuri Gagarin, 1961) into orbit around the Earth. They also sent the first female and black cosmonauts (Russia's equivalent of astronauts) into space (Valentina Tereshkova in 1963 and Arnaldo Tamayo Méndez in 1980). They took photographs of the Moon's 'dark side' before anyone else did (1959) and were the first to bring samples of lunar soil back to Earth by probe (1970). They built the first ever space station (*Salyut 1*, 1971), and performed the first ever spacewalk (Alexey Leonov in 1965). They were the first nation to land on any world other than Earth (*Luna 2* performed a hard landing, or crash, on the Moon in 1959, and *Luna 9* a soft landing in 1966), and send back photographs (*Luna 9*, 1966). They sent probes to Venus (the *Venera* probes, from 1961 onwards) and Mars (*Mars 1*, in 1962), which attempted to fly past, photograph and land on both planets (sadly, these were only partially successful). The US may have won the Space Race, but the USSR dominated it.

This intense focus on spaceflight was reflected in the popular media of the time. Many societies have dabbled in science fiction, Mars being a very common topic, and the USSR was no exception. In 1923, author Aleksey Nikolayevich Tolstoy delivered one of the very first Soviet science-fiction novels in the form of *Aelita*, which was developed into a silent film of the same name the following year (commonly dubbed the first ever Soviet science-fiction film, and one of the earliest feature-length films worldwide to focus on space travel and Mars).

Aelita tells the story of two Russians (an engineer and a soldier) travelling to Mars, where they meet a fractured Martian civilisation attempting to survive on a dying planet. The protagonist falls in love with the princess of Mars, the eponymous Aelita, before having to flee back to Earth to escape a burgeoning social rebellion (driven by the large gap and wealth disparity between the upper and working classes – obviously drawing on the political theme of capitalism run riot and a reflection of the USSR's fears at the time).

The associated film tweaked the story somewhat but kept some of the same themes. It was very successful and the team involved with promoting the film pulled out all the stops. In the film, the protagonist detects bizarre and cryptic radio signals that are ostensibly from Mars. 'Anta ... Odeli ... Uta,' they repeat, 'Anta ... Odeli ... Uta.' Various Soviet newspapers printed this cryptic phrase in the lead-up to the first showing of *Aelita,* claiming that the signals had been heard worldwide and finally deciphered! Any interested parties would find out the meaning, they said, on the film's release date at the cinema. There was apparently such crowding and demand for tickets on the evening of the film's premiere that the director himself, Yakov Protazanov, was unable to attend.

While the USSR reached Mars fever particularly early, the rest of the world wasn't far behind. The US, for example, hit a similar phase just after World War II, in which the country was gripped by science fiction and an eager anticipation for the technological achievements to come.

In the 1950s, American education on space travel came from an unlikely source: Disney. The animation company produced three episodes of its 'Disneyland' television series themed around space: 'Man in Space', 'Man and the Moon' and 'Mars and Beyond'. All three episodes were based on a famous series of 1950s magazine articles called 'Man Will Conquer Space Soon!'.

The scientific advisor for the episodes was Wernher von Braun. That name may well leap out as familiar: von Braun was a prominent German rocket scientist during World War II and later worked for the US space programme, where he was chief architect for the rocket-booster system that enabled the Apollo missions to get to the Moon. It's thought that both Walt Disney and von Braun were keen to channel public interest in science fiction into real-life science – they wanted to harness public enthusiasm and use it to actually get to space (even today, some scientists state that getting to Mars is essentially a matter of money and politics, so von Braun and Disney were on the right track). The way to do this, they thought, was via television – and the timing was certainly right! The decades following the late 1940s are commonly dubbed 'the Golden Age of Television'.

'To make people believe that spaceflight was a possibility was [von Braun's] greatest accomplishment,' said Mike Wright, a staff historian for NASA's Marshall Space Flight Center, in 2002. 'Von Braun brought all of this out of the realm of science fiction.' In an interview in 1965, Disney also elaborated on his reasoning for tackling space exploration in his films. 'If I can help through my TV shows … to wake people up to the fact we've got to keep exploring, I'll do it.'

All three space-themed 'Disneyland' episodes are as enjoyable as expected, with quirky and fantastical animations of bobbling robots, whizzing planets and blazing rocket ships, all set to a wonderfully nostalgic staccato voiceover. However, they were also surprisingly informative. Walt himself introduces each of the episodes, sometimes dubiously aided by a robot co-host named Garco. The episode dedicated to Mars tackles issues such as the human history of astronomical observations, Mars in science fiction, UFOs, the formation of the Solar System, the emergence of life, interplanetary travel and more, all accompanied by classic Disney animation.

The episode also plays around with several of the different preconceptions about Mars floating around Europe during the Age of Enlightenment (which began in the mid-late 1600s). Describing the writings of French author and academic Bernard Le Bovier de Fontenelle (1657–1757), Disney animated Mars as being a rusty red colour, with bizarre, lanky, two-stilted creatures, each with a single eye and long beak, hopping around and shooting beams of light from their pupils. Despite his imaginings, de Fontenelle was not a particular lover of Mars; in his 1686 book *Conversations on the Plurality of Worlds*, he delves into each planet in individual detail and is pretty dismissive of the Red Planet.

'I don't know that there is any thing remarkable in this planet,' he writes. 'The days there are about half an hour longer than ours, and the years twice the length of ours, except a month and a half. Mars is four times less than the Earth, and the Sun appears rather smaller and less brilliant than it does to us. In short, Mars contains nothing calculated to arrest our attention.'

Moving on to share a Swedish perspective, Disney's Mars animation transforms into a dark, black scene, with two floating orbs surrounded by a buzzing mass of tangled wires and threads. The accompanying voiceover describes the creatures as fibre-covered beings that communicate by means of telepathy.

The episode also touches on various fictional representations of Mars, from Kurd Lasswitz's depiction of large-eyed humanoid Martians surviving on synthetic food (1897), to Robert Braine's dream (1892) of mute but technologically advanced 10-foot-tall music-loving Martians who are able to communicate with Earth via a telescope-bearing plant, to the more familiar portraits shown in H. G. Wells's *The War of the Worlds* (1898) and Edgar Rice Burroughs's *Barsoom* series (1912 onwards). (More on Mars in science fiction later.)

H. G. Wells did not just entertain ideas of Mars through his fiction, he also wrote a more serious article about a decade after the publication of his most famous title for *Cosmopolitan* magazine, in which he described 'the flora and fauna of our neighbouring planet, based upon scientific reasoning' (1908). He concluded that Martian flora – green and leafy, showing 'the bluish green of a springtime pine' – would be 'slenderer and finer and the texture of the plant itself laxer' than plants here on Earth, due to Mars's reduced gravity. The typical Martian plant, said Wells, is far taller than one on Earth, with 'bunches and clusters of spiky bluish green leaves upon uplifting reedy stalks'. There would be no fish on Mars, as life that relies on underwater breathing or gills 'must have been exterminated … long ago', nor 'flies nor sparrows nor dogs nor cats – but we shall probably find a sort of insect life fluttering high amidst the vegetation'. The Martians themselves, he postulated, will probably be large (between two and three times heavier and taller) humanoids with backbones, eyes, large brains, 'grotesquely caricaturing humanity'; they may even be covered in feathers or fur.

'How wild and extravagant all this reads!' wrote Wells. 'One tries to picture feather-covered men nine or ten feet tall, with proboscides and several feet, and one feels a kind of disgust of the imagination. Yet wild and extravagant as these dim visions of unseen creatures may seem, it is logic and ascertained fact that forces us towards the belief that some such creatures are living now. And, after all, has the reader ever looked at a cow and tried to imagine how it would feel to come upon such a creature with its knobs and horns and queer projections suddenly for the first time?'

An Earth-shattering 'kaboom'!

Disney and von Braun were among the first to popularise spaceflight to the American public. Some have even claimed that the series helped to stir up such popular support that it essentially secured years of funding for the country's space programme.

Even today, nowhere is Mars more inescapable than on television. While some shows only superficially reference Mars – *Veronica Mars*, for example, which appears to have only used the planet for naming inspiration, and the UK series of *Life on Mars*, which has some sci-fi and cosmic themes but doesn't draw heavily from Mars itself – many others have dedicated episodes, characters, or entire storylines to Mars, or explicitly referenced our attempts to reach the Red Planet. *My Favourite Martian, Biker Mice from Mars, Captain Scarlet and the Mysterons, The Simpsons, Sesame Street, South Park, The Muppets, Looney Tunes, Doctor Who, Futurama, Pinky and the Brain, Animaniacs, Dragonball Z, Star Trek, Flash Gordon* – the list goes on.

Creating perhaps one of the most famous Martian characters of all time, Warner Bros (*Looney Tunes* and *Merrie Melodies*) introduced the hapless Marvin in 1948, a shadowy little humanoid/ant-like thing with a green skirt and plumed helmet (like the Roman god Mars) and a relentless, desperate desire to destroy Earth. This was usually more self-centred than malicious – Marvin only tries to blow up Earth because it blocks his view of Venus. Marvin was apparently created as a mirror for the gun-toting and moustachioed Yosemite Sam;

where Sam was loud, brash and ineffective, Marvin was quiet and unassuming, yet had the potential to be truly destructive. Marvin's nefarious plots were usually thwarted by Bugs Bunny, leaving him wailing and distressed, lamenting the lack of an 'Earth-shattering "kaboom"'!

Alongside a number of starring roles in his own animated shorts, Marvin and his faithful robotic canine companion K-9 often popped up in other animated universes, including those of *The Simpsons, South Park, Futurama, Pinky and the Brain, Animaniacs* and other Warner Bros shows and spin-offs, including *Space Jam* (in which he was the game's referee), *Taz-Mania* and *The Sylvester & Tweety Mysteries*. He attempted to reshape the Solar System using the wonderfully named *Illudium Q-36 Explosive Space Modulator* and carried a futuristic ray gun.

Bugs may have flippantly likened Marvin to a spittoon-topped bowling ball, but Marvin's character and design have been pretty influential. NASA's *Spirit* rover, for example, adopted Marvin as its official mascot, choosing a launch patch depicting Marvin enthusiastically saluting in front of a red, white, and blue background bearing the words 'Red Planet Gladiators'! In a 2001 episode of the BBC's now-defunct *Omnibus* series, Steven Spielberg joked that he wouldn't be surprised if George Lucas eventually admitted that he'd drawn inspiration from Marvin in creating Darth Vader's entire iconic suit and mask. Between 2008 and 2011 there were talks of a Marvin the Martian live-action animated film potentially starring Mike Myers, but there have been no updates since. We may never know if this is a blessing or a curse.

Martians as a whole have served as inspiration for many. As well as being a nostalgic trip down memory lane, exploring each instance tells us much about the societal and scientific attitudes at the time. One recurring theme that strongly defines Mars, and has since the year dot, is water – either the abundance or lack of it.

The British television series *Doctor Who*, for example, drew heavily on the idea of water for their multiple Martian species, some of which were developed in the 1960s (when the possibility of extant and developed Martian life was just about dying out). Despite having an entire fictional universe to play with, *Doctor Who* has used Mars as a home planet for several different species, namely the Gandorans, the Flood and the Ice Warriors.

The Flood was the collective name for a hive-mind virus that existed within and spread through water – the species starred in a *Doctor Who* special episode set in 2059 (entitled *The Waters of Mars*), in which it managed to infect the water supply of the appropriately named Bowie Base One colony. The residents of the base were humans that had relocated from Earth; after eating food washed in infected water due to a broken water filter, the residents of the base began showing symptoms including spasms, convulsions, cracked skin, black teeth, icy blue eyes, the ability to survive unscathed on the surface of Mars and – most prominently – a constant flood of water streaming from their skin, hands and mouth. To stop The Flood from managing to leave Mars and spread to other planets, the Tenth Doctor (David Tennant) intervened and blew up the base, leaving just a stark warning behind urging future visitors to avoid the water.

The Ice Warriors first appeared in 1967 (to Patrick Troughton's Doctor), and have starred and been mentioned several times since – most recently in a 2013 episode of the revived series (to Matt Smith's Doctor) entitled 'Cold War'. An ancient and long-living race of cold-blooded reptilian humanoid cyborgs (quite a mouthful!), the Ice Warriors looked fearsome, but were sometimes willing to cooperate with the Doctor. They stood up to 7ft tall with scaly armour, fangs and clawed hands; strongly preferring cold climates, they could be defeated by extreme heat and were created, or evolved from, the Gandorans, the native inhabitants of Mars who initially ruled over them. The recurring theme throughout their appearances is ice and snow. The Doctor speculates on how they may have frozen themselves within ice on Mars or its moon Deimos to preserve their species. At the height of their civilisation, the Warriors were peaceful and happily existed on Mars alongside water-filled canals, but things deteriorated over time. In a way, the Ice Warriors resemble the 'Face on Mars', a naturally occurring rock form ('mesa') that resembled a face in low-resolution photos of Mars taken in 1976 (after the birth of the Warriors – coincidence, or conspiracy?).

Other television series airing at the same time as the older *Doctor Who* episodes decided to instead highlight the mysteriousness of potential Martian life. At the time, we had no clue what lurked on Mars, if anything – something that Gerry and Sylvia Anderson, creators of the popular 1967 *Captain Scarlet and the Mysterons* series, used to their advantage.

The Andersons conceived the idea for *Captain Scarlet* in the 1960s. Set in 2068 – worryingly soon! – and themed around the concept of an interplanetary war raging between Earth and Mars, *Captain Scarlet* was markedly less light-hearted than the Andersons' previous work, which included the hugely successful *Thunderbirds* series and *Stingray*. Gerry Anderson coined the portmanteau 'supermarionation' to describe his iconic and quirky style of mixing marionette puppets with special effects and scaled-down model sets in his animations.

The Mysterons were a race of warriors from Mars unlike many other fictional alien races that had been depicted before. Rather than resembling standard life forms, bizarre aliens or tentacled monsters like H. G. Wells's *The War of the Worlds* antagonists or the comedic cauliflower brains of *Mars Attacks!*, the Mysterons were computer-based beings that banded together as a collective consciousness. Using their unique power of 'retrometabolism', Mysterons could manipulate matter to create obedient facsimiles of whoever they liked – including Captain Scarlet, who was killed and recreated by the Mysterons in the series's very first episode. After escaping Mysteron control, Captain Scarlet became a key character in protecting Earth in the Mars-Earth war. He was an expert pilot, astronaut and James-Bond-like sleuth with devilish courage and a cutting wit. He retained a number of Mysteron characteristics, including rapid healing (retrometabolism) and the ability to sense any nearby Mysterons.

While the concept is a great one in itself, it tells us much about the time period in which *Captain Scarlet* was born. In the 1960s, speculation about life on Mars was everywhere. It might seem very recent, but astronomers still knew little about whether or not Mars was inhabited – and, if it was, by whom. Astronomers gazing at the planet saw things that suggested the planet might not be a completely empty world, from vast, dusty looking plains to dark patches that seemed to grow and subside with the seasons, reminiscent of vegetation, to growing and shrinking polar caps, as on Earth, to the most outlandish suggestions of weird criss-crossing artificial canals and radio bursts. Mars was still a very alien place.

When asked about how he dreamt up the Mysterons, Anderson attributed it to the prevalence of theories about life on Mars flying around in the 1960s. 'I thought we should make a show about the Martians,' he said in a 2002 interview with *Billboard* magazine, 'but then there were doubts being expressed by scientists as to whether the so-called "canals" on Mars were really man-made. Since we were well into pre-production, I came up with the idea of making the Martians invisible, so if they did come up with conclusive evidence that there was no life on Mars, I could say, "Ha-ha, yes there is – but you can't see it".'

The mix of excitement and uncertainty around Martian life gave creators a huge space to work within. DC Comics introduced the Martian Manhunter (J'onn J'onzz) in 1955, an original member of the Justice League (along with his better-known colleagues Batman, Superman, Wonder Woman, the Flash, Green Lantern and Aquaman – he also starred in the TV spin-offs *Smallville* and *Supergirl*). He is a muscular humanoid Green Martian, which is exactly what it says on the tin: a green-skinned human-like being from Mars. He wears a blue cloak and boots, and a couple of crossed red belts across his chest.

The Martian Manhunter suffers from the same kind of 'problem' that Superman has in that he simply has too many powers. He is highly intelligent, can read minds, move objects with his mind, become invisible, fly, quickly heal himself and regenerate missing body parts, and shape-shift (including growing to a huge size, stretching his limbs and even altering the structure of his cells so he can pass through solid objects). He has superhuman strength and is incredibly resistant to injury, hunger, thirst, suffocation – pretty much any form of discomfort, as he can simply change his physical form to adapt to his environment – has X-ray vision and can shoot lasers from his eyes. His psychic abilities are some of the most powerful in the entire DC universe; he can manipulate others' thoughts, plant false memories and control his enemies' minds.

His only weakness? Fire! Flames scare the Manhunter so much that his mind loses control, causing him to melt down into plasma and lose his physical form – a satisfyingly mundane kryptonite for such an accomplished being. The character may not be included in the upcoming Justice League film (*Justice League*, 2017); a prominent DC screenwriter, David Goyer (*Man of Steel, The Dark Knight* trilogy) was less than enthusiastic about the idea, calling him 'goofy' and a less-than-popular character.

Films are similarly stuffed with Mars-related content. Tellingly, there are too many to name; some of the better-known offerings include *The War of the Worlds, Invaders from Mars, Mars Attacks!, Total Recall, My Favourite Martian, Red Planet, Mission to Mars, Ghosts of Mars, Doom* (based on the popular video game), *Mars Needs Moms* and *John Carter.*

Some of these are better than others. *Mars Attacks!* was released in 1996 and has since achieved acclaim as a so-bad-it's-brilliant cult classic – although it's still not considered a good film by many. It was intended as a spoof or parody of a sci-fi B-movie, but in some ways fell just on the wrong side of satire. As renowned film critic Roger Ebert phrased it, '*Mars Attacks!* is not so much a comedy as a replica of tacky old saucer movies – not so much a parody as the real thing. What's wrong with this movie [is that] the makers felt superior to the material. To be funny, even schlock has to believe in itself.'

The film was directed by Tim Burton and starred a pretty impressive line-up: Jack Nicholson, Glenn Close, Annette Bening, Pierce Brosnan, Danny DeVito, Michael J. Fox, Sarah Jessica Parker, Natalie Portman, Jack Black, even Tom Jones. It depicted Martians as techno-babbling, cauliflower-brained, gun-toting aggressors, intent on decimating humanity with their superior weaponry and penchant for destruction. Spoiler alert: the Martians are eventually defeated when a character discovers their weakness to Earth music (when Slim Whitman's 'Indian Love Call' is played, their brains explode in a gruesome fashion).

The idea for *Mars Attacks!* was not plucked from thin air. It was based on a series of Topps trading cards released in the 1960s. This 55-piece series was very popular with children (but less so with their parents). The artwork was gory and violent, telling the story of an Earth invaded and overrun by cruel Martians – many of the scenes were actually adapted for and used in Burton's film, and the same cauliflower-brained illustrations featured heavily. Burning sheep, sizzling flesh, immolated humans, ominous flying saucers, kidnapped women being consumed by giant mutant spiders, toppling skyscrapers, panicked crowds being swept up and squished into buildings, planes cut in half mid-flight, dogs being zapped by extraterrestrial laser guns (in front of their owners, some of whom are tearful children), Martian brains being bludgeoned and torn apart, Martians performing scientific experiments on (conscious) humans while their loved ones watch on, beheadings, even a child innocently clutching a doll while offering a Martian an ice cream (ominously titled 'Last Licks').

Many parents were shocked by the content of the cards and the number of complaints abruptly ended the set's production soon after it went on sale. As with many collectible crazes, these card sets are now quite valuable and a full set in good condition can fetch thousands of pounds. Following the initial cards, the 'Mars Attacks' franchise continued with a series of comic books, numerous re-issues and developments of the original trading card series (most recently in 2015), plush toys, table-top games, costumes, books, clothing and other merchandise. In May 2015, Topps announced their plans to re-release a new 72-card series called 'Mars Attacks: Occupation'. They went about this in a distinctly modern way, asking for funds on crowdfunding platform Kickstarter and promising 'amazing pulpy art … horrific scenes of alien mayhem, terrifying atrocities, over-the-top madness, and of course a healthy dose of blood and babes'. The project received nearly US$200,000 of funding, far surpassing the original goal of $50,000, proving that 'Mars Attacks' fever is still alive and well over half a century later.

The success of the franchise demonstrates one of Mars's greatest appeals: the planet offers an accessible and somewhat-known-but-somewhat-mysterious setting for all kinds of imaginative storylines. For this reason, video games love using Mars-related maps or themes – colonisation, space travel, dying and dystopian societies, scientific research settlements gone wrong, cosmic war, aliens, the unknown.

First-person shooter game *Red Faction* features a miner based on Mars who rebels against an authoritative corporation. The upcoming game *Lacuna Passage* is entirely based on a protagonist attempting to survive and explore Mars, while searching for her missing crewmates from a manned mission to the planet (excitingly, it will use actual imagery from NASA's *Mars Reconnaissance Orbiter* in the gameplay!). *Doom* depicts a research outpost located on the Red Planet whose occupants must battle demons swarming on to Mars from hell, transported by teleportation portals located on Mars's moons (Phobos and Deimos). Other games, including *Destiny* and *Wipeout*, have unlockable Martian maps and stages – if you can find them, that is. The more light-hearted *Daffy Duck: The Marvin Missions* uses the aforementioned ant-like humanoid Marvin the Martian as the antagonist, who must be destroyed in order to complete the final stage of the game.

New concepts are being developed using virtual reality (VR) technology. NASA is collaborating with game developers on a project named *Mars 2030*, an interactive game designed for VR headsets that will place the player directly on to the surface of Mars. *Mars 2030* is as scientifically accurate as possible, from topography through to gravity, making it a kind of hybrid between a video game and a planetary simulator.

Many musicians have taken inspiration from Mars, notably Holst with his famous *The Planets* suite (written in approximately 1914). Holst took more of an astrological stance on the planets, describing each by what he deemed to be its leading characteristic. Venus was the Bringer of Peace, Mercury the Winged Messenger, Jupiter the Bringer of Jollity, Saturn the Bringer of Old Age, Uranus was the Magician and Neptune the Mystic. Mars was, predictably, the Bringer of War. The Holst Museum describes the piece as conveying 'raw Martian impulses: the mis-use of the will, the desire for action and the chaotic energy of rebellious youth'.

Several more modern acts have used the planet in their name – Bruno Mars wanted something 'out of this world' for his moniker, The Mars Volta were inspired by their fascination for science fiction and 30 Seconds to Mars's Jared Leto explained the name as a 'reference, a metaphor for the future', also liking the idea that it refers to something, a 30-second trip to Mars, that's 'so close [but] not a tangible idea'. Mars had perhaps the biggest influence on David Bowie. Bowie pops up time and time again relating to Mars; after the musician's death in 2016, an online petition was even started to rename the planet after him, garnering thousands of supporters!

First up was Bowie's obvious ode to Mars, 'Life on Mars?', a track that featured on his 1971 album *Hunky Dory*. While it obviously referred to the Red Planet, Bowie used Mars more symbolically than literally; the song tracks the thoughts of a mousy-haired girl who is growing increasingly disenfranchised and unhappy with both her reality and the media she consumes. She wonders and hopes if there is possibly a better life out there somewhere – perhaps life on Mars? – and sadly longs to be a part of it.

Bowie then created an entire persona, Ziggy Stardust, based on the idea of a Martian living on Earth. The eponymous album, *The Rise and Fall of Ziggy Stardust and the Spiders from Mars*, was released in 1972 and is consistently ranked as one of the best albums of all time.

Originally intended as a concept album (which 'kind of got broken up, because I found other songs I wanted to put in the album which wouldn't have fit into the story,' Bowie reportedly said), *Ziggy Stardust* still manages to lead the listener through a complete narrative.

In the words of Bowie himself, Ziggy is a 'Martian messiah who twanged a guitar', a washed-up rock 'n' roll star whose music is no longer relevant to Earth's youth. At the beginning of the album ('Five Years'), the protagonist is despondent and downcast, aware that Earth is just five years from an apocalyptic death. However, he begins to sing songs of hope and possibility ('Starman') that resonate with the despairing inhabitants of Earth. He gains a global following and becomes a megastar along with his backing musicians, the Spiders from Mars. His music is actually influenced by extraterrestrial beings, the so-called 'starmen', who visit and advise him in dreams. However, Ziggy begins to believe his own hype, considering himself a prophet, a god, and his arrogance leads to his eventual demise when the starmen and his own fans tear him apart on stage ('Rock 'n' Roll Suicide').

Bowie has elaborated on various parts of this story at times, calling the starmen 'infinites, black-hole jumpers … They really are a black hole, but I've made them people because it would be very hard to explain a black hole on stage'. The infinites land very specifically in Greenwich Village and take parts of Ziggy's physical form in order to build their own bodies, as 'in their original state they are anti-matter and cannot exist on our world'.

However, in true Bowie fashion, this fantastical storyline is just one of the musician's interpretations of his own album. 'It depends in which state you listen to it,' he said in an interview. 'The times that I've listened to it, I've had a number of meanings out of the album, but I always do … I find that I learn a lot from my own albums about me.'

Mr Fix-It and the man cave

Current beliefs about the Red Planet aren't so different to those held throughout history. Mars is still viewed as a masculine, aggressive presence. Take, for example, the wildly successful *Men are from Mars, Women are from Venus*, a self-help book published in the 1990s. Marketed as the definitive guide to sustaining a successful and happy relationship, the book sold millions of copies and became the highest-ranked non-fiction work of the decade.

In the book, author John Gray asserts that the genders are ruled by norms and desires so different that they can be considered to be from entirely separate planets. It's as if Martians and Venusians both decided to come and settle on Earth, but promptly forgot their different origins and became baffled by their inevitable incompatibility.

Men (Martians) are uncommunicative and thrive on proving themselves, valuing their autonomy, power and competence above all else. They assume the role of 'Mr Fix-It', usually offering solutions over comfort; 'while women fantasise about romance, men fantasise about powerful cars, faster computers, gadgets, gizmos, and new powerful technology'. Women (Venusians), on the other hand, value 'love, communication, beauty and relationships'; they are nurturing, supportive, with an interest in personal growth and spirituality. As Gray puts it, 'on Venus, everyone studies psychology and has a master's degree in counselling'. When stressed, men retreat to their Martian caves, while women instinctively need to talk through what's on their mind. Martians and Venusians keep score of the 'acts of love' – from doing the washing-up to offering a hug to buying a bouquet of flowers – they give and receive in their relationship in different ways, which can lead to conflict. The book even includes a 'Martian/Venusian Phrase Dictionary', which unpicks incomprehensible phrases such as 'I'm fine' and 'It's OK', to help you understand what on Earth your alien of a partner is blabbering on about.

While the book is primarily intended as a tool to facilitate better communication (debatable) and doesn't claim to be scientifically precise (thankfully), the approach has been criticised heavily for perpetuating – and strengthening – societal stereotypes about femininity and masculinity. Numerous studies and publications have disagreed with *Men are from Mars, Women are from Venus,* stating that it dangerously encourages dismissal of one's feelings, and justification of one's poor communication, on the basis of gender alone. Gray's philosophy can be seen as an overly simplistic play on the traditional astrological and mythological attributes for the character of Mars: that he is stoutly masculine and capable, a one-dimensional representation of power and physicality, ambition and action (and equally of Venus, a planet named after the goddess of love and beauty that represents sympathy, empathy, unity, harmony, love and the arts).

The idea of Mars and Venus representing separate worlds of thinking (although in a genderless capacity) has also been popularised by American historian and policy specialist Robert Kagan, who commented that 'Americans are from Mars, Europeans are from Venus' when considering power, militarism, and foreign policy. While Europeans seek a 'self-contained world of laws and rules and transnational negotiation and cooperation', Americans believe that 'true security and the defence and promotion of a liberal order still depend on the possession and use of military might'.

Kagan expanded on this view in his bestselling essay *Of Paradise and Power: America and Europe in the New World Order*, published in 2003. 'On major strategic and international questions today, Americans are from Mars and Europeans are from Venus,' wrote Kagan. 'They agree on little and understand one another less and less.'

CHAPTER FOUR

Death Stars and Little Green Martians

In his original series of *Cosmos*, the inimitable Carl Sagan dedicated an entire episode to Mars. 'Martians! Why so many speculations and fantasies about Martians, rather than Saturnians, say, or Plutonians?' he asked, as fuzzy 1980s-style renderings of red spheres swept into view.

He had a point. When you imagine Mars, what springs to mind?

It may be a dry and barren red landscape populated only by trundling robot visitors from Earth, occasionally using their hi-tech cameras to take pictures of their lonely surroundings. Or an alien colonisation, with sleek black obelisks and giant tin cans on long, wiry legs, firing off capsules into space to invade their nearest neighbours. Maybe it's little green men – wobbly, muscular or humanoid, with gangly limbs, futuristic helmets or three huge eyes, leaving slug-like trails of sticky goop in their wake.

It's likely that your answer reflects the amount and type of science fiction you've watched or read over the years. From the aforementioned Marvin the Martian and black comedy of *Mars Attacks!* to the imagined worlds of H. G. Wells and Arthur C. Clarke, Mars pops up all over the place and has done for years. If you were to picture the planets Venus or Mercury, your mental image is likely to be quite different and far less detailed.

Tellingly, and with more than a touch of irony, Sagan didn't dedicate any other *Cosmos* episodes to a single planet. He did, however, go on to answer his own question.

We love Mars because it seems to be very Earth-like, said Sagan in the aforementioned episode of *Cosmos* ('Blues for a Red Planet', October 1980). 'It's the nearest planet whose surface we can see,' he explained as he bobbed in front of the camera in his trademark polo jumper and beige blazer. 'There are polar ice caps, drifting white clouds, raging dust storms, seasonally changing patterns, even a 24-hour day. It's tempting to think of it as an inhabited world.'

Like Earth, Mars is a bit like a big, rocky onion ('differentiated', in astronomer-speak) with different layers of material all wrapped around a metallic core. The thick outermost surface layer – the crust – is covered by a layer of vivid red-orange dust and soil. It has a thin atmosphere that's not all that similar to Earth's, admittedly, but far closer than Mercury's tiny tenuous one or Venus's thick smog. It has volcanoes, cliffs and craters scattered across its surface, and is orbited by a couple of small moons. It is around half the diameter of Earth, but has about the same amount of land area because Earth is so covered in water. The planet's axis is slightly tilted, meaning that Mars experiences seasons similar to those on Earth – albeit more severe, due to Mars's more elliptical orbit, and just under twice as long, as it takes Mars 687 days to zip once around the Sun.

Going on this (very) simplified factsheet, the idea of life on Mars doesn't seem all that far-fetched. Astronomers in the late 1800s and early 1900s agreed. Well into the twentieth century the idea of life on Mars, and intelligent life at that, was believed to be a very real possibility.

The *canali* saga

Early 'Mars fever' may have been triggered by a single astronomical mistake … pun intended.

In 1877 Mars hit opposition, lining up in space with Earth and the Sun – aligned Sun–Earth–Mars – and nearing its closest approach to our planet. Opposition makes a planet brighter, bigger, and more easily observable throughout the day and night. For Mars this happens roughly every couple of years, but the 1877 opposition was particularly important – not only did it see astronomer Asaph Hall's discovery of both Martian moons, Phobos and Deimos, it also saw the birth of an idea that was to capture the imagination of the public, other astronomers and the science-fiction community for decades.

In September 1877, Italian astronomer Giovanni Schiaparelli was trying to construct the most detailed ever map of Mars. As he peered at the planet's surface he spotted something intriguing – a criss-crossing maze of long, straight grooves covering the Martian surface. He dubbed these *canali*, or 'channels', and described them as quite shallow depressions in the Martian soil that extended in a 'straight direction for thousands of miles, over a width of 100, 200 kilometres and maybe more'.

'In the absence of rain on Mars,' Schiaparelli wrote, 'these channels are probably the main mechanism by which the water, and with it organic life, can spread on the dry surface of the planet.'

Here came the astronomical mistake. Although nothing in Schiaparelli's observations suggested that these grooves themselves might have been artificially made, *canali* appeared in English translations as 'canals', not 'channels'. A subtle change, but enough to shift the perception of Schiaparelli's *canali* from something natural to something intelligently designed and manufactured.

There could only be one explanation for this grid of artificial alien canals: Mars was a world that had once hosted – or was still hosting? – a civilisation of intelligent, living, breathing aliens. After all, rivers on Earth were meandering, not straight. As archaeologist Charlie Holloway (Logan Marshall-Green) says in the 2012 Ridley Scott film *Prometheus*, 'God does not build in straight lines'. (This is a paraphrasing of author Gertrude Stein's apparent assertion that there are no straight lines in nature. We now know that it's not true – just look at crystals, snowflakes and honeycomb.)

Before this, there had been other bits and pieces indicating that some familiar features may exist on Mars. Mars seemed to look different at different times of the year, suggesting that the planet experienced the same kind of seasonal shifts that we do here on Earth: things like melting and freezing ice caps, snowdrifts, and vegetation and forests slowly changing colour.

Astronomer Percival Lowell excitedly linked these things to Schiaparelli's canals in the years following the 1877 opposition. Lowell was a respected astronomer whose calculations led to the successful discovery of Pluto some years after his death; the former planet's name was chosen partially because its first two letters are Lowell's initials.

According to Lowell, Mars had water ice caps that melted every spring and re-froze every winter. In spring the water from these polar regions was carried down towards the drier equatorial regions of Mars via artificial channels and used by the Martians to grow crops. These crops accounted for the shifting patterns of dark and light that danced across the surface of the planet and changed throughout the year (now known to be caused by a mixture of winds and seasonal phenomena, such as defrosting ice within surface sand). Intelligent technologically and politically advanced Martians were desperately trying to sustain life on the surface of a drying and dying planet.

Lowell met with a mix of scepticism and support. Although some astronomers backed up his and Schiaparelli's observations in the decade that followed, others were simply unable to spot an intricate system of canals – or, sometimes, any canal-like features at all. Regardless, Lowell gained a reputation for being a Mars expert, writing several books on the planet that described it as a cool, arid world perfectly capable of supporting advanced life. He became absolutely certain that Mars was home to beings of some sort or other, despite his contemporaries continually making new findings that suggested he might be wrong.

Respected scientist Alfred Russel Wallace, for example – co-'discoverer' of Darwinian evolution by natural selection – argued that Martian air would be much too cold and thin for liquid water to exist, and pointed out that spectroscopic studies of Mars hadn't managed to find any evidence of water. He was quite outspoken about his finding, exclaiming that 'only a race of madmen would build canals under such conditions'!

Conversely, Schiaparelli himself was actually somewhat supportive of the idea of intelligent Martians with a penchant for canal-building, so there may be a whiff of myth to the *canali* saga. He certainly made no effort to correct the *canali*/canals 'mistranslation', and went on to work closely with Lowell for the remainder of his life, putting together continually updated maps of Mars.

Even if the translation wasn't a mistake, this single word triggered a wave of Mars-themed science fiction. Some of the more prominent works that dreamt up a canal-covered Mars include astronomer and author Camille Flammarion's 1889 novel *Uranie* (*Urania* in later English versions), Tarzan creator Edgar Rice Burroughs's 1912 *Barsoom* series, Robert A. Heinlein's 1949 *Red Planet*, parts of Ray Bradbury's *The Martian Chronicles* in the 1950s and, later on, Kim Stanley Robinson's 1990s *Mars* trilogy. The canals in these novels ranged from thin irrigation canals like those found on Earth, to trenches filled with frozen water, to channels burnt into the Martian soil by scorching sunlight.

Incidentally, the *Barsoom* series, namely the book *A Princess of Mars* starring a protagonist named John Carter of Mars, was recently adapted into a 2012 Disney film named *John Carter,* removing all reference to Mars in the title. Rather than advertising and emphasising the story's association with Mars, the filmmakers strove to play it down. The film's director, Andrew Stanton, was quoted as saying, 'Here's the real truth of it. I'd already changed it from *A Princess of Mars* to *John Carter of Mars.* I don't like to get fixated on it, but I changed *Princess of Mars* ... because not a single boy would go. And then the other truth is, no girl would go to see *John Carter of Mars.*'

This is a small, but interesting, reflection of how mainstream media perspectives on science-fiction films are slowly changing. While Disney executives decided Mars was a stigma and too weak a selling point for either gender in 2012, this seems like an odd marketing tactic when you consider the more recent success of films like *The Martian* (2015), which couldn't be more proudly Mars-themed if it tried! Despite

Stanton agonising over the name of the film it was still a colossal box-office failure, losing Disney tens (or even hundreds) of millions of dollars. As Disney optimistically phrased it, '[the film] failed to connect with audiences as much as we had all hoped'.

This title swap incensed British film critic Mark Kermode, who went on a characteristic rant about the film in 'Kermode and Mayo's Film Review', a podcast produced by the UK's BBC Radio 5 Live featuring Kermode and his colleague Simon Mayo. Kermode declared the adjusted title to be 'rubbish … an entirely negative thing', but placed more blame on the overall plot and production.

'It's incomprehensible, boring, turgid, dull, plodding, really risible, awful, with a thrangy-throng-in-the-hiddly-bang-diddly-bong-bong script – the kind of stuff that makes George Lucas's opening screen at the beginning of *The Phantom Menace* really sound like Shakespearean sonnets or Chaucerian verse or T. S. Eliot's *The Wasteland*,' said Kermode. 'At the end of it, it's boring, it's really boring. It's an ocean liner, an oil tanker, a behemoth, a huge great articulated lorry of a movie, lumbering from set piece to set piece, with people explaining the plot to you – "with the fourth frang of the fiddly fong, from the planet fruskar frlong" – it's boring, and expensive, I mean really … *boring*.'

'I got that,' replied Mayo drily.

Eventually, Schiaparelli and Lowell's canals were shown not to exist. Photographs sent back from NASA's *Mariner* fly-bys of Mars in the 1960s spotted no canal system to speak of. Although Mars does have deep rifts, channels and trenches scattered across its surface, they are natural geological features. The intricate system of canals is now generally thought to have been an optical illusion, caused by the human brain's tendency to search for patterns and straight lines. Lowell also reported spoke-like features on Venus that were later shown not to exist, suggesting that he may have been seeing shadows cast on his retina by the blood vessels in his eye.

Canals aside, the suggestion of a dry, arid Mars that was slowly dying and shrivelling up was a powerful image that lasted much longer than the idea of canals. One notable story was penned by Walter Tevis in 1963: *The Man Who Fell to Earth.* The 1976 film, now a cult classic despite middling reviews, details the story of an advanced humanoid Martian named Thomas Newton, played by David Bowie. Newton comes to Earth in search of water, as his home planet is rapidly drying up. He uses his superior intellect to patent a number of inventions and accrue significant wealth, which he uses to build a spacecraft. Before he can leave Earth, however, he is seized by the suspicious authorities and imprisoned, and slowly becomes addicted and ruined by the many unfamiliar vices of Earth (namely, alcohol and television).

Bowie is also linked to the 1961 Robert A. Heinlein novel *Stranger in a Strange Land*, whose plot bears remarkable similarities to not only *The Man Who Fell to Earth* but also that of the concept album *The Rise and Fall of Ziggy Stardust and the Spiders from Mars.* As mentioned previously, both the *Ziggy Stardust* story and Heinlein's novel star an androgynous alien living in a near-apocalyptic world who gains a following, becomes almost god-like, and is eventually attacked and killed by an angry mob. Although Bowie was no stranger to using science-fiction inspiration in his music, lyrics and various on-stage personas, he wasn't a fan of Heinlein's book, saying in a 1974 *Rolling Stone* interview, 'I don't like the book much. In fact, I think it's terrible. It was suggested to me that I make it into a movie, then I got around to reading it. It seemed a bit too flower-powery and that made me a bit wary.'

H. G. Wells's *The War of the Worlds* (1898), arguably one of the most influential sci-fi novels ever written, is also based around the idea of a dry Mars. In the story, Earth is attacked by mysterious capsules fired from a giant space gun on the surface of Mars. The capsules slowly unscrew to reveal heaving, pulsating, bear-sized Martians with grey-brown tentacles and lipless mouths. This is one view of Martians as being completely unlike anything we find here on Earth: 'inhuman, crippled and monstrous', Wells adds. These

Martians go on to climb into the iconic giant three-legged machines that are equipped with lethal heat rays, and use humans for food.

While Wells doesn't mention canals, his Martians attack because they need our water. They're thirsty and unable to sustain themselves on their parched home planet.

The War of the Worlds is a first-person narrative that portrays the plot as a series of real events happening to the poor narrator. For Halloween in 1938 the book was adapted for American radio. It was narrated and directed by actor and filmmaker Orson Welles, and read out in the style of a series of news bulletins as if a Martian attack was currently taking place. This style was unusual for the radio series, and caused a lot of panic and confusion from listeners tuning in at different times and missing the show's context. The media reported that 'the country at large was almost frightened out of its wits. Men called radio stations offering to enlist, others were panic-stricken'.

Two years after this broadcast, Orson Welles met H. G. Wells in San Antonio, Texas, US, and the two men chatted on the radio. After jokingly mocking the spelling of Welles's surname, Wells asked whether there actually had been widespread panic in the US following the broadcast … 'Or wasn't it pure Halloween fun?'

'[It was] the same kind of excitement we extract from a practical joke in which someone puts a sheet over their head and says "boo!",' said Welles. 'I don't think that anybody believes that that individual is a ghost, but we do scream and yell and rush down the hall. And that's just about what happened.'

Welles's radio broadcast may have just been 'Halloween fun', but other adaptations were somewhat more sinister. In February of 1949, an Ecuadorian radio station decided to air their own version of *The War of the Worlds,* in a similar format to Welles's deceptive fake news bulletins. Listeners became agitated and panicked, some even taking refuge in nearby church buildings. Once the radio station's dramatic director, Leonardo Páez, realised what was happening he stopped the broadcast and issued an apology, which only turned the panic to fury. By the end of the night, angry mobs had attacked the radio premises, set them alight and up to 20 people had died – and Páez had been arrested.

From the Big Three to Bradbury Landing

Gulliver's Travels is a staple of many childhoods. The story is perhaps best remembered by the iconic scene in which Gulliver is captured and tied down by a race of tiny six-inch-tall people. Because of this, the book might not be immediately placed in your 'sci-fi' box, but it actually contains an intriguing scientific prediction.

After a series of fantastic adventures, Gulliver encounters a levitating island known as Laputa, which is inhabited by a civilisation obsessed with mathematics and science. As an interesting side note, Swift's account of Laputa is a thinly veiled critique of the abstract science that his society was experiencing at the time: despite being great fans of mathematics, the Laputians couldn't apply it practically to save their lives. When Gulliver arrives on Laputa, a tailor attempts to measure him for a suit. Rather than using a flexible tape measure, he instead uses a quadrant, ruler and compasses to figure out Gulliver's dimensions, resulting in a suit that was 'very ill made, and quite out of shape'.

More importantly for our focus on Mars, Swift writes that the astronomy-loving Laputians had discovered 'two lesser stars, or satellites, which revolve about Mars', which raced around the planet in 10 hours and 21.5 hours respectively.

Gulliver's Travels was written in 1726, over 150 years before the discovery of Mars's moons, Phobos and Deimos. Phobos, the larger inner moon, takes 8 hours to zip once around Mars, while Deimos lags behind at just over 30 hours.

While he wasn't bang on, Swift was pretty close! It's likely that he was influenced by what we knew of the Solar System at the time: Earth had one moon and Jupiter had four known moons. Sitting between the two, it was likely that Mars would have at least two. It's sadly unlikely that Swift had psychic powers (and yes, this was actually suggested at the time), but he undoubtedly made an impressive guess.

Later authors also dreamed up Martian moons, but all attempts to find any in reality failed until Asaph Hall's discovery during the aforementioned 1877 opposition. When Hall was trying to spot Phobos and Deimos he worked with seeing conditions so awful that he almost gave up completely. However, he persevered and eventually managed to spot the elusive moons that scientists had previously failed to find in significantly better weather. It's likely that Hall looked closer into Mars than others had – or simply just got lucky.

Some had a more unusual solution to this puzzle. In 1959, American college professor Walter Scott Houston suggested that Phobos and Deimos might actually be artificial satellites! He published this claim, attributing it to 'Dr Arthur Hayall of the University of the Sierras', in the monthly newspaper of his astronomical society … on 1 April. April Fool! Houston later explained that he chose the idea as it was 'so ludicrous it would not need to be labelled a gag'.

However, others didn't take the theory as an obvious joke. Just one month later, Russian scientist Iosif Shklovsky reportedly spoke of Mars's moons perhaps being artificial, declaring that 'no natural causes can explain the origin of the Martian moons'. Although his statements seemed to be genuine, Shklovsky's peers scoffed at the idea that the scientist truly believed the theory, instead suggesting that he 'must have said that with tongue in cheek, just to see what newspapers would make of it … he is much too brilliant to believe such nonsense'.

Some years later, plant physiologist Frank Salisbury wrote a paper that appeared in the January 1967 issue of the journal *Bio-Science*. In it he explored the idea of extraterrestrial life and UFO sightings, discussing how Mars's moons were small and had incredibly circular and well-behaved orbits – in other words, they 'have certain of the characteristics of artificial satellites,' he wrote. Salisbury was loath to discount the idea of Phobos and Deimos being Martian spaceships, and encouraged scientists to be open-minded about the hypothesis (although he did consider some of the opposing arguments to be 'quite impressive').

The idea of a couple of Martian Death Stars might sound far-fetched today – and even at the time was dubbed 'science-fiction nonsense' by many – but this idea held at least some traction in the 1960s, reflecting just how late this way of thinking persisted. Soon after this, NASA's *Mariner* (Mars) missions and the later 1970s *Viking* probes reduced our evidence for an inhabited Mars, and the idea of a technologically advanced and canal-building civilisation slowly faded.

In his *Cosmos* episode, Sagan calls Mars 'a kind of mythic arena on to which we've projected our Earthly hopes and fears'. As with Swift's abstract Laputians, many authors latched on to the idea of a Martian civilisation to critique aspects of Earth they disliked. In a way all fiction is guilty of this, but Mars provided a whole other mysterious society to play with.

Many of the big names in science fiction have written about the Red Planet: John Wyndham, Philip K. Dick, Kurt Vonnegut, C. S. Lewis, Robert A. Heinlein, Arthur C. Clarke, H. G. Wells, Isaac Asimov, John W. Campbell, Stanley G. Weinbaum, Ray Bradbury ... the list goes on. Heinlein, Clarke and Asimov are often referred to as the 'Big Three' – even if you're not really a sci-fi fan, you'll likely have heard of them.

Clarke, author of *2001: A Space Odyssey*, made his first foray into science fiction with *The Sands of Mars* in 1951. In the book humans attempt to grow crops on Mars, and find small kangaroo-like creatures there. The researchers hope to ignite Phobos and use it as a second heat and light source for Mars, to prepare for an eventual human colonisation of the planet. NASA's *2001 Mars Odyssey* orbiter, currently in orbit around Mars, was named in honour of Clarke.

Heinlein produced a particularly famous story in the aforementioned *Stranger in a Strange Land*, which stars a human named Valentine Michael Smith who was born on a human expedition to Mars and raised by Martians: a kind of cosmic version of *The Jungle Book*. He comes to Earth and integrates into terrestrial culture, becoming something of a celebrity on Earth due to his air of baffled wonder, psychic abilities and superhuman intelligence. The book was massively successful upon its release in 1961; it was described as 'the most famous science-fiction novel ever written', was included as one of 88 'Books that Shaped America' in 2012 and received the 1962 Hugo Award for Best Novel.

Rounding off the 'Big Three' is Isaac Asimov. As with Clarke's orbiter, Asimov has both an asteroid and a Martian crater named in his honour. Asimov wrote *The Martian Way* in 1952, a description of the colonisation of the Solar System and disagreements between the different colonies. In Asimov's tale an Earth-based politician decides that settlements on Mars, Venus and the Moon should no longer be supplied with water shipments, as they've become leeches on terrestrial society. The Mars colony instead attempts to steal a chunk of Saturn's rings, which are chock-full of water. They succeed and end up offering to sell some of their water supply to the remaining colonies in place of Earth. Asimov's book was politically motivated and aimed to criticise the political repression and fear-mongering ways of 1950s McCarthyism in America. Asimov was wary of publishing the book, expecting a strong backlash, but actually experienced nothing at all. He was reportedly quite irritated by this, noting that 'I must have been too subtle – or too unimportant'.

Many other authors got in on the Mars action. Kim Stanley Robinson's more recent *Mars* trilogy (*Red Mars, Green Mars, Blue Mars*) is incredibly comprehensive, chronicling two centuries of colonisation and terraformation of Mars. The trilogy creates an entirely new universe and has been called the 'gold standard' of realistic science-fiction writing (in recent years, *The Martian* has garnered similar praise). In Robinson's books, the titular adjectives relate to the state of the planet at the time; it starts pre-colonisation, when the planet is dusty and arid and red, turns green during a period of terraforming (the converting of Mars into a more

Earth-like world), and ends up blue after all the water locked up in Mars's soil and polar caps is able to melt and survive on the adapted surface.

Robinson dreamt up a thriving society on the Red Planet, even going so far as to design a flag for the 'Martian' population. Robinson's flag comprised, predictably, strips of red, green and blue. Other fictional flags include one proposed in the 1993 science-fiction novel *Moving Mars:* a flag cut in half across the diagonal, with the upper half coloured royal blue and featuring three red dots (a larger one for Mars and two smaller for its moons). Heinlein's *Stranger in a Strange Land* depicted a far simpler flag – pure white with the symbol for Mars, a circle with an arrow pointing to the upper right, centred in red. There is actually no official flag for the planet; any colonisation attempt would be subject to the laws of the Outer Space Treaty, a document signed in 1967, which does not allow any country or nation to lay claim to any part of any celestial body. The Earth does not yet have an agreed-upon planetary flag, either, although those of the United Nations and of Earth Day (a photograph of Earth on a navy background) are the most prominent contenders.

But this is skipping ahead a little; it's worth briefly going back to the very beginning. One of the first books to dream up a mysterious Mars popped up in 1889: *Melbourne and Mars: My Mysterious Life on Two Planets*, by Australian writer John Fraser. This held a more mystical focus than the Martian invasion stories that soon followed.

Melbourne and Mars presents itself as the edited diaries of a merchant who begins experiencing night visions. These visions are actually a telepathic link to a child called Charlie: his 'other self' on a utopian Mars. Fraser paints Mars as having very advanced technology, something he believed to be the key driving force for growth and development here on Earth. Although it's tricky to dig up much information about Fraser, it seems that not all of his books were in the same vein. Before penning *Melbourne and Mars*, Fraser wrote the dubiously titled *How to Read Men as Open Books* and

Husbands: How to Select Them, How to Manage Them, How to Keep Them. Lovely!

A modern author who played around with mentally advanced Martians was Ray Bradbury, author of the well-known *Fahrenheit 451*. Bradbury got very creative with Mars in his book *The Martian Chronicles*. One chapter was previously published under the title *Mars is Heaven!* in 1948, and describes the third attempted voyage to Mars by explorers from Earth. When they arrive, the crew look out of their ship's window and spot a small town just like one back home on Earth. The crew's captain, John Black, is understandably hesitant about leaving the ship and describes his misgivings, warily noting the bizarre similarities with Earth (specialised plant life, cupolas, porch swings, pianos, 'a piece of music titled, strangely enough, "Beautiful Ohio" … which means that we have an Ohio River here on Mars!').

Sadly, Captain John Black and archaeologist Samuel Hinkston met a sticky end. Leaving the ship, they found the small Earth-like town to be full of relatives and old friends, some of whom had died years earlier. After letting their guard down and joyfully joining the Martians, Black again feels uneasy and the idea correctly strikes him that this may not be quite as innocent as he first thought. He correctly surmises that the Martians might be manipulating his memories and emotions to evade suspicion and plot a malicious attack to get rid of him and his crew … as had happened to the previous two Earth missions to Mars, both of which disappeared without a trace.

Mars is Heaven! appeared as a chapter named 'The Third Expedition' in Bradbury's *Martian Chronicles* in 1950. Bradbury was an incredibly influential name in science fiction, and such a lover of Mars in his fiction that the team behind NASA's Mars *Curiosity* rover named the rover's landing site 'Bradbury Landing'. Bradbury liked the idea of Mars as a mirror of Earth rather than 'a seer's crystal in which to read a miraculous future', imagining explorers finding a 'somewhat shopworn image of themselves' on the Red Planet.

As with the aforementioned *John Carter* disaster, *The War of the Worlds* and *Aelita*, some of these stories ended up being adapted for the big screen. It's worth mentioning Philip K. Dick here; Dick's fiction has inspired and produced an impressive number of hit films, including *Blade Runner, Total Recall* and *Minority Report*. Dick was unsatisfied with our world, and so tried to critique it and build a better one in his fiction. *Total Recall* was based on Dick's 1966 short story *We Can Remember It for You Wholesale*, in which an ordinary man from a technologically advanced Earth wants to visit Mars, but can't afford it. He visits a company to buy implanted false memories to simulate the trip. However, he begins to remember that he actually has previously visited the Red Planet as a secret government assassin and had his real memories wiped. The film is perhaps best known for its iconic Martian mutants, three-breasted prostitutes and Arnie's eyes almost popping out of his head.

There are too many works to list individually, showing just how far-reaching Mars's influence has been for over two centuries. Just the few stories mentioned tackle the ideas of colonisation and industrialisation of the Solar System, Martians and Earthlings visiting and invading one another, terraforming and sustaining life on Mars, Martians with telepathy, hypnosis and mental manipulation, the realness of reality, memory, politics, technology and more.

NASA's *Phoenix* lander, which reached the Martian surface in May 2008, paid homage to the many different human imaginations of Mars that undoubtedly helped it get there. On board was a DVD entitled *Visions of Mars*. This disc held numerous pieces of fiction and non-fiction from a whole host of authors including Carl Sagan, Isaac Asimov, Ray Bradbury, Kim Stanley Robinson, Arthur C. Clarke, Percival Lowell and many more, accompanied by visual artwork. The very first Martian library!

'These tales and images have inspired generations about the wonder of space, including many men and women who are now researchers and engineers in the space programme,' explained Louis Friedman, executive director of the Planetary Society, who conceived the idea for *Visions of Mars*. Many others have echoed these sentiments through the years. Sagan, for example, apparently once said that his unwavering love of space – and Mars in particular – was sparked by an early love of sci-fi from authors like H. G. Wells and Edgar Rice Burroughs. In turn, Sagan has inspired the next generation of science communicators, including Neil deGrasse Tyson, an astronomer who continued Sagan's *Cosmos: A Personal Voyage* legacy with 2014's *Cosmos: A Spacetime Odyssey*.

Both runs of *Cosmos* have been massively successful, but not all of the media released about Mars has been so influential or well regarded. One piece of Mars-themed fiction is often referred to as one of the worst films ever made – a 1964 film called *Santa Claus Conquers the Martians*. A reviewer at *Empire* magazine didn't hold back on what they thought of the festive film. 'For those of you who haven't watched this movie – and if so, lucky you – this is a film so shockingly shit that it almost transcends its shitness and enters the ethereal plane of so-shit-it's-really-really-good. Alas, *Santa Claus Conquers the Martians* is so good at being shit that it falls out the back end of the so-shit-it's-really-really-good category and becomes really, really shit again.'

The Martians in the film – named things like Stropo and Dropo – look suspiciously like humans dressed up as aliens, with skin-tight green lycra bodysuits, small television antennae spiking out of the top of their heads and a murky green-brown skin tone. Still, we're meant to believe that these aliens are very technologically advanced. They kidnap Santa in a spaceship and have small robot helpers. 'Their robots look like your little cousin walking around with a bin on his head,' ranted *Empire*. 'What. A. Movie.'

What would Mark Watney do?

It would be impossible to write about Mars's influence on sci-fi without mentioning *The Martian*. A hugely successful film released in 2015, *The Martian* was based on a 2011 novel of the same name by Andy Weir. It ranked as the tenth highest-grossing film in 2015, with a worldwide gross of over US$630 million (as of August 2016, making it Ridley's highest-grossing film to date) and, unlike *Santa Claus Conquers the Martians,* received hugely positive reviews across the board.

The Martian is different to many of the other sci-fi titles mentioned in this chapter in that it focuses more on the 'science' than the 'fiction' – it aimed to be as accurate as possible, and those working on the film have dubbed it 'as much science fact as science fiction'.

Rather than expanding on existing Mars tropes, the film instead played off more familiar and general themes, such as the abandoned hero, the idea of an inexplicably talented main character (sometimes called a 'Mary Sue' or 'Marty Stu'), the big rescue, distrust of authority, romance, man versus nature and more.

As a brief plot summary, *The Martian* describes a manned mission to the Red Planet. The team are out on the surface of Mars dutifully collecting samples when they're alerted about an incoming dust storm, which threatens to topple their spacecraft and leave them stranded. The astronauts quickly head back to the ship, but are forced to leave one team member, Mark Watney (Matt Damon), behind when he is struck by storm debris and (mistakenly) presumed dead.

The subsequent two hours document Watney's struggle to survive. Watney has an acerbic sense of humour, vast reserves of ingenuity and technical skill, and a passion for botany, which he ends up using to grow enough food – potatoes, which have one of the highest calorie-per-acre yields of any staple crop – to survive until he can be rescued. There are multiple little nods and references throughout the film for people who love space exploration, most notably when Watney needs to locate, fix and use the communications system from the defunct *Mars Pathfinder* probe to speak to NASA on Earth (last operational in the late 1990s).

Web comic xkcd perfectly summarised the appeal of *The Martian* in June 2015 (comic 1536): 'You know the scene in

Apollo 13 where the guy says "we have to figure out how to connect *this* thing to *this* thing using *this* table full of parts or the astronauts will all die?" *The Martian* is for people who wish the whole movie had just been more of that scene.'

The Martian is an excellent example of factually correct science fiction. Although not a scientist himself, Weir was very conscientious about ensuring his story was as accurate as possible, while taking a few artistic liberties for the sake of storytelling. 'I spent about half my time researching,' Weir has said. 'I'm not worried about my work coming under the eye of experts. It's a work of fiction, and it's going to have inaccuracies. That's just how it is. But getting a thumbs up from those experts and hearing them say "it's more accurate than anything else in the field" is a good feeling.'

The movie took a similar approach, taking an important step forward in the sci-fi arena. Rather than ignoring science in favour of overblown CGI explosions and wacky-looking aliens, the production team behind *The Martian* collaborated with NASA like never before – one of the best (if not *the* best) box-office crossovers between science and science fiction yet. In an interview with *Wired* magazine, Weir explained that NASA really liked his book and saw the associated film 'as an opportunity to re-engage the public with space travel'. British physicist and science communicator Brian Cox also saw the practical appeal of *The Martian,* calling it 'the best advert for a career in engineering I've ever seen'.

This has indeed proved to be an effective tactic, so much so that a writer at the *Washington Post* even (boldly) claimed that 'Andy Weir and his book *The Martian* may have saved NASA and the entire space program'. Although that specific claim is likely to have more than a hint of clickbait about it, NASA has certainly used the publicity from *The Martian* very effectively, even coordinating their scientific press releases with big release dates and movie deadlines, and releasing web articles and interactive tools that plot Watney's fictional journey across the surface of Mars. In an article for *Astronomy* magazine, one of the scientific consultants on the film, NASA's Jim Green, claimed this to be a concerted effort to reach '"the Mars generation" – millennials who've never seen humans leave low Earth orbit'.

'Hopefully the message in a movie like *The Martian* is one that really galvanises participation in stuff like [space exploration] and makes people excited about science,' said Matt Damon in a press interview for *The Martian*. 'I mean the writer of the movie, when I first met with him, he said, "I see this as a love letter to science, that's really what I want to make". We talked a lot about it and said, "Yeah, that's a cool thing to put out into the world right now".'

CHAPTER FIVE

The Motions of Mars

Ancient astronomers struggled to figure out the motions of Mars. For many centuries they scratched their heads trying to figure out its orbit, pondering over why the planet's weird dance through the sky just didn't seem to fit with what they thought to be true – and why making it fit was so incredibly difficult.

Until the mid-1500s, a few things were widely believed to hold true throughout the cosmos. One, all the planets moved through space in perfect circles. Two, they all whizzed around at constant speeds. Three, all the planets orbited in exactly the same plane, which slices through the centre of the Earth. Accompanying these main three assumptions was a fourth, namely that all celestial bodies were formed from unchanging material with constant properties: the heavens were immutable and eternal.

Mars was a major fly in the ointment. Its orbit was strange. In fact, Mars doesn't move around the Sun in a circle, but in a giant ellipse, something we now know to be true for all planets. However, Mars's ellipse is unexpectedly eccentric. Eccentricity can be thought of as a measure of how 'squashed' a circle is, if you were to press down on it or pinch its sides. An orbit's eccentricity is measured on a scale from zero to one, zero being a perfect circle and one an open-ended parabola (not a confined orbit, more like a giant arc).

Long-period comets, for example, whoosh into our inner Solar System from way out beyond the orbit of Neptune and form a giant curved path around the Sun. Their orbits are so stretched and eccentric that some of them are actually open-ended, meaning that they will never loop back around and we'll never enjoy their company again.

Earth's orbit, on the other hand, is pretty circular, with an eccentricity of just 0.017. The award for most circular planetary orbit in the Solar System goes to our sister planet Venus, which has a tiny eccentricity of 0.007.

With a value of 0.094, Mars's orbit is more eccentric than that of any other planet in our Solar System bar Mercury, which beats it with a value of 0.206. Its distance from the Sun varies significantly; at the innermost point of its orbit Mars lies 207 million km from the Sun (129 million miles or 1.38 AU, where 1 AU is the Earth–Sun distance), but at its most distant it moves out to nearly 250 million km away (155 million miles or 1.67 AU). Overall, it sits an average of 228 million km from the Sun (142 million miles or 1.52 AU).

This eccentricity causes the planet to vary in brightness (it can be tens of times brighter at some points than at others), colour (from ochre-yellow through to blood-red) and size (depending on its ever-changing distance from us) with each new orbit, something that didn't fit in with the eternal, symmetric, predictable heavens that early astronomers were picturing and attempting to explain.

Mars is also a key player in the story of retrograde motion, as mentioned previously. As the planet is so close to us and obvious in our skies compared with its more distant planetary siblings, Mars's weird zigzagging movements are better known and more visible in the sky. In fact, all planets in our Solar System periodically behave this way, or would do if they were easily visible to the naked eye.

For these reasons and more, many astronomers have struggled to compile an accurate view of the cosmos since antiquity.

Many ancient civilisations – the Babylonians, Egyptians, Chinese – have tracked the planet through the sky, noting its red colour and weird motions. However, Aristotle (*c.*384–322 BC), a philosopher and scholar dubbed 'the first genuine scientist in history' by *Encyclopaedia Britannica*, was one of the first known to attempt to compile a comprehensive model of planetary dynamics. Aristotle was a student of Plato (*c.*427–347 BC), and together the two are thought to have formed the basis for modern-day academia (philosophy and science in particular).

Astronomers in Aristotle's day shared many beliefs that no longer hold true. Aristotle believed there were four or five main elements, namely water, fire, air, earth and aether (otherwise known as 'quintessence', essentially the 'stuff' that fills space), that the Earth was the centre of the Universe, that planets moved in perfect circles around the Earth, and more. At the time, astronomers also believed the stars to be constant, unmoving presences: fixed, unmoving studs in the distance.

Aristotle viewed the cosmos – which at the time was just the Solar System – as a set of fixed crystalline spheres. He believed that the stars and planets were all stuck within a Russian-doll-like set of concentric thin orbs, nestled within one another. His universe comprised a number of these shells suspended in space, each containing a planet, all rotating. Due to the then-unquestionable belief that the Earth sat at the centre of the Universe, Aristotle's spheres were ordered to explain the various motions he observed in the skies around him. The Earth sat at the centre, held within a set of spheres containing (in order of increasing distance) the Moon, Mercury, Venus, the Sun, Mars, Jupiter, Saturn and the stars. A large 'prime mover' shell encapsulated the entire system and moved round at a constant angular velocity, causing all the inner shells to rotate.

To match his model with his observations, Aristotle played around with the velocities of the individual shells, but was unable to explain why the planets varied in brightness, or the aforementioned zigzag of retrograde motion. According to his model, each planet remained at a fixed distance from Earth, held fast by its crystal shell, and moved at a constant speed.

Aristotle was actually responsible for observing one of the first recorded lunar occultations of Mars. On 4 May 357 BC, Aristotle watched as the Moon passed in front of Mars in the night sky. The event indicated that the planets were further away than our moon and helped Aristotle to arrange his model of the heavens.

Later astronomers, including Aristarchus of Samos (*c.*310–230 BC)*, Apollonius of Perga (*c.*262–190 BC) and, most famously, Ptolemy (*c.*90–170 AD), tried to fix this problem. Some came across what they deemed the perfect solution – add more spheres!

Building on Aristotle's Universe as a base, astronomers introduced the idea of 'epicycles'. Picture a single circular line representing a planetary orbit, known in this model as a 'deferent', with a single dot on it. Rather than representing a planet, this dot acts as the centre of yet another, smaller, circular line, this time known as an 'epicycle'. The planet itself sits on and moves around this epicycle, as the central dot moves along the deferent. In the Ptolemaic Universe, all planets moved on these small epicyclic circles, which in turn latched on to their deferent orbits around the Earth (not the Sun). The entire Solar System was a dizzying dance of circle upon circle upon circle.

While contrived, these epicycles did fix many of the gaps in Aristotle's theory. They led to planetary motions resembling helical coils around the central body, in this case the Earth, and altered the distance between any given planet and the Earth at different times in orbit. As such, epicycles could be used to explain phenomena such as Mars's retrograde motion and varying planetary brightness without compromising on the core principles of the time (constant speeds, perfect circles, geocentrism).

Ptolemy built even further on epicycles … by adding even more spheres (yes, really). In some particularly tricky cases, an epicycle itself might have an epicycle of its own, totalling

---

* While an Earth-centric (geocentric) Universe was accepted by the vast majority of astronomers, some disagreed. Aristarchus of Samos was one of the first to suggest a heliocentric Universe, but the controversial idea didn't gain traction until many hundreds of years later. The history of astronomy is peppered with frustrating instances like this, of scientists arriving at a correct idea only to be dismissed by the scientific community (or dismissing it themselves as too outlandish).

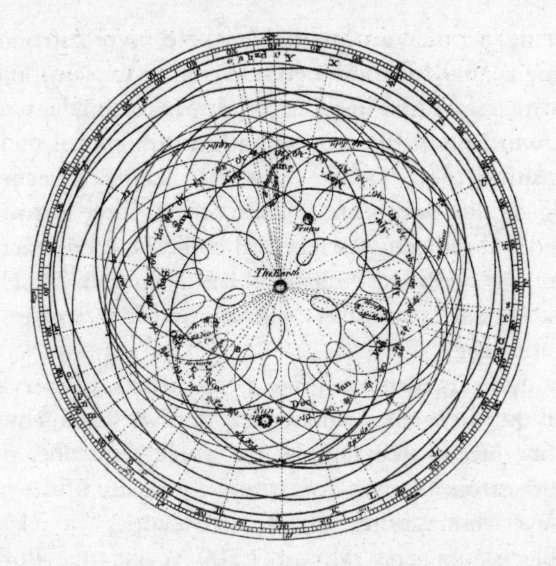

An 18th-century illustration of the geocentric Solar System, showing the Sun, Mercury, and Venus whirling on epicycles around the Earth.

three independently whirling circles. To account for other phenomena that still couldn't be explained by that model, Ptolemy also altered the Earth's position in the model, shifting it slightly from the centre of each deferent, and stated that the planets actually assumed perfect circular motion around an 'equant', a point displaced to the other side of the deferent's centre by the same distance as Earth. Both Ptolemy and Aristotle described Mars's retrograde motion as real motion – Mars was *actually* moving backwards – and not apparent, as we now know to be true.

As an interesting note, the phrase 'adding epicycles' has entered popular use and is used to describe cases where someone clings to a theory beyond all reason, attempting to force it to match our observations. It's often used in reference to fields such as homeopathy, astrology or creationism, and other pseudoscientific ideas. It's the philosophical opposite to Occam's razor, which states that the simpler a theory and the fewer its assumptions, the better (although this is, ironically, an overly simplistic explanation of the principle).

Ptolemy himself accepted that there were astronomical phenomena that his model couldn't quite explain, and that there were contradictions in his work – for example, wouldn't planets whirling madly around within thin crystal shells end up smashing their shell? – but was happy to accept an 'instrumentalist' approach, as Plato did. As long as his theory appeared to be mostly correct and could be used practically, Ptolemy was satisfied. (Ptolemy did also concern himself with the realist side of things, but more often spoke as an instrumentalist.)

Although this may seem like a case of astronomers trying to alter the shape of a square peg until it sort-of-but-not-really-fits into a circular hole, it's worth remembering that pre-1600 astronomy was done using a mixture of theory and naked-eye observations. The first telescopes (or telescope-like objects) appeared almost 1,500 years after Ptolemy's death. It was also incredibly challenging – and dangerous – to oppose the religious thinking of the time. The most famous example of this is 'the Galileo affair', a sequence of events beginning in the 1610s in which Galileo was condemned and ultimately imprisoned by the Catholic church for promoting heliocentrism (a Sun-centric Universe).

Religious rule didn't only affect Galileo. Nicolaus Copernicus (1473–1543), an astronomer who lived roughly a century before Galileo and is another of modern astronomy's big names, was nearly stymied by his fear of encountering such a backlash.

By Copernicus's time, the gaps in Ptolemaic thinking were becoming increasingly problematic. Some centuries earlier, Arab scientist Ibn al-Haytham (*c.*965–1040 AD) had attacked Ptolemy's view, taking particular offence to the former's instrumentalist perspective. Al-Haytham dubbed the Ptolemaic Universe 'an arrangement that cannot exist … the fact that this arrangement produces in his imagination the motions that belong to the planets does not free him from the error he committed in his assumed arrangement. For a man to imagine a circle in the heavens, and to imagine the planet moving in it, does not bring about the planet's motion'. He may not have been polite, but nor was he wrong!

Copernicus saw similar problems. As he approached the end of his life, he tried to align the old theories of Ptolemy and Aristotle with what he saw as he studied the sky. In the year of his death, 1543, Copernicus published his famous tome *De Revolutionibus Orbium Coelestium (On the Revolutions of the Heavenly Spheres)*, in which he suggested the revolutionary idea that the Earth was not at the centre of the known Universe – the Sun was. He also suggested that the Earth moved in not one but two ways, both moving through space around the Sun and spinning on its axis.

His assistant Rheticus, a scientist from present-day Austria best known for his work in trigonometry, helped him to publish *De Revolutionibus*; although Copernicus had formulated his ideas many years earlier, he had been so worried about a negative response from both the scientific and religious communities that he buried his work and decided not to publish it. Rheticus disagreed. He persuaded his teacher to publish the work, and took on editing duties as he prepared the manuscript.

While Copernicus clung to a number of older ideas – mostly motivated by his fear of abandoning certain philosophical views – his new theory was a huge leap forward for science and triggered a huge change in astronomical thinking. Rather than a geocentric system, we now had one that placed the Sun at the centre of the Solar System, with Earth only forming the gravitational centre of the Earth–Moon pairing. The stars and sky did not rotate around the Earth; instead the Earth rotated on its axis, giving rise to the apparent motion of the cosmos.

Before publishing, Rheticus went through Copernicus's writings in detail, focusing heavily on the section concerning Mars. Mars still didn't quite fit. Copernicus still claimed that planets moved on fixed and perfectly circular orbits around the Sun, which didn't quite explain Mars's behaviour in the sky. He also agreed with many facets of Aristotelian and Ptolemaic physics, namely the existence of Russian-doll-like shells and complex epicycles (although it's possible that Copernicus did not believe in solid shells made of crystal, but orbs that were either symbolic or more fluid).

Copernicus's description of Mars infuriated and puzzled Rheticus. *De Revolutionibus* required multiple complex epicycles to explain the planet's motion. No matter how hard Rheticus tried to simplify this, it didn't seem to work. Legend has it that one day he became so frustrated that he begged the spirits to help him. An apparition appeared and violently threw Rheticus about, smashing his head first on the ceiling and then on the floor, with a parting roar: 'These are the motions of Mars!'

The Mars story involves a great many of the leading names and figures in the history of astronomy. Mars's motions also preoccupied Danish astronomer Tycho Brahe and his German colleague Johannes Kepler, both active in the 1500s and 1600s.

Brahe's career began a few years after Copernicus's death. Despite his work still taking place in a pre-telescopic era, Brahe made strikingly accurate observations of Mars and its position in the sky. These data were later used by Kepler to prove that Mars didn't orbit on a circular path, but an elliptical one. Although Brahe's theories and ideas were not always correct, he was revered for gathering meticulous and incredibly accurate observations. Early in his life, he stated that this was his very aim.

'I've studied all available charts of the planets and stars and none of them match the others,' he noted in 1563 (at the age of just 17; researching Brahe is a great way to make one feel inadequate). 'There are just as many measurements and methods as there are astronomers and all of them disagree. What's needed is a long-term project with the aim of mapping the heavens conducted from a single location over several years.'

Although Brahe wasn't able to come up with a theory that worked fully, he worked on several key concepts that eventually led to a heliocentric breakthrough. Brahe compiled all his thoughts into his own theory of the Universe and produced a geo-heliocentric model, in which the Moon and Sun orbited the Earth and all other planets orbited the Sun.

In a way, Brahe's model was an intermediary between Ptolemy and Copernicus – it maintained the Earth's importance, removed the solid planetary shells but kept circular orbits, scrapped the immutable stars and heavens, and kept the stars relatively close to the outermost planet (at the time, Saturn).

In the late 1500s, Brahe met Johannes Kepler (1571–1630), who subsequently became his assistant. Brahe tasked Kepler with understanding Mars's orbit, a task that was readily accepted. Sadly, Brahe died in 1601 before he could see his assistant's conclusions, but Kepler continued his work in earnest and some eight years later published his *Astronomia Nova* (*New Astronomy*). Kepler's big breakthrough was to picture planetary orbits as ellipses, not circles, with the Sun at one of the foci (essentially the 'centres' of the ellipse – a circle has just one focus, its centre, but ellipses have two due to their elongated shape along one axis). He also figured out that the planets didn't move at constant speeds: they moved faster as they approached the Sun and slower when further away.

Kepler relied heavily on Brahe's observations of Mars to develop his three famous laws of planetary motion: that the orbit of each planet is an ellipse with the Sun at one of the foci, that each planet sweeps out equal areas in space in equal time and that the square of a body's orbital period is proportional to the cube of its mean distance from the Sun. The idea that planetary orbits were not perfectly circular was a huge leap forward in planetary theory and in understanding the motions of Mars.

Hail, children of Mars!

Mars has been a key player in astronomy for millennia, due in large part to its proximity to Earth. While the more distant outer planets – Uranus, Neptune, the now-demoted Pluto – were only discovered following the invention of the telescope in the 1600s, the inner and nearby outer planets have always been visible with the naked eye, and so have been included as part of our Solar System models since the year dot.

The invention of the telescope enabled us to characterise our nearest neighbours better than ever before. Many astronomers turned their eyes Mars-ward; Galileo Galilei, widely regarded as one of the fathers of the telescope, is thought to have observed the planet from 1608 to 1610, completing some of the very first telescopic views of Mars. These views were not particularly detailed, nor were they formally recorded, but Galileo did tentatively discuss his findings with his peers. He observed Mars to subtly change in diameter as it moved through its orbit, something that was not possible in the Ptolemaic system. 'I ought not to claim that I can see the phases of Mars; however, unless I am deceiving myself, I believe I have seen that it is not perfectly round,' he wrote.

Galileo was observing the planet's phases. As Mars is a superior planet, meaning that it moves round the Sun outside the path of Earth's orbit, it cannot show a crescent phase, but it can show a gibbous one (essentially the opposite of a crescent phase, with most of the body illuminated rather than blacked out). Regardless of its position in orbit, from Earth around 85 per cent of the planet is well-lit at all times. In order for us to see the face of Mars as dark, the planet would have to move between Earth and the Sun. As this can't happen we never see a new or crescent Mars, only gibbous or full.

In 1610, Galileo discovered four of Jupiter's moons: Io, Europa, Ganymede and Callisto, known as the 'Galilean moons'. This finding was another nail in the coffin of geocentrism: if Earth was the centre of the Universe, how could moons orbit another body? Later that year Kepler wrote to Galileo and stated his wonder at the finding, speculating on how it might apply to the rest of the Solar System. 'I should rather wish that I now had a telescope at hand,' he wrote, 'with which I might anticipate you in discovering two satellites of Mars (as the relationship seems to me to require) and six or eight satellites of Saturn, with one each perhaps for Venus and Mercury.'

Kepler was absolutely correct about the size of Mars's little family. Mars's two moons, Phobos and Deimos, were not discovered until 1877, but Kepler's proposition was inadvertently kept alive by Galileo. In the same year, Galileo noticed something unsettling about Saturn – through a telescope, it appeared to be a composite of three orbs, not one (this optical effect was later shown to be due to Saturn's extensive ring system, which was a completely alien phenomenon at the time). Galileo found himself in a sticky situation: he was unwilling to announce his discovery without a published paper, but wanted to ensure that others knew he had been the first to make the finding on the off-chance that someone else were to stumble upon it in the meantime. He decided to send a cryptic letter to some of his contemporaries, including Kepler, in August 1610, containing a Latin anagram:

s m a i s m r m i l m e p o e t a l e u m i b u n e n u g t t a u i r a s

Although Galileo intended the message to be decoded as '*Altissimum planetam tergeminum observavi*', translated as 'I have observed the most distant of the planets to have a triple form', Kepler demonstrated the issues inherent in communicating via anagram, instead translating it – in a remarkable coincidence – as '*Salue umbistineum geminatum Martia proles*', or 'Hail, twin companionship, children of Mars'! Kindly, Galileo didn't leave Kepler in suspense for long. Later that year he cleared up the mystery, declaring his real finding to the astronomical community.

It took later observations – by astronomers including Francisco Fontana in the 1630s, Christiaan Huygens and Giovanni Cassini in the mid-late 1600s, Giacomo F. Maraldi between 1672 and 1719, and others – to glean more about Mars and its properties.

Fontana produced some of the first known drawings of Mars in the 1630s, showing that Mars's disc was uneven in colour. Huygens followed up a couple of decades later, creating some of the most detailed astronomical drawings that had ever been made of the Red Planet. Between them, the separate observations from Huygens and Cassini revealed multiple features on the Martian surface, including patchy regions of dark and light, a large dark spot – now known as Syrtis Major – and white-hued polar caps.

Syrtis Major is the most obvious dark patch ('albedo feature') observable on the face of Mars as it rotates. It was later discovered to be a low-relief shield volcano by NASA's *Mars Global Surveyor*, and was the first surface feature ever to be discovered and documented on the surface of another planet. Huygens watched the dark patch move across the face of Mars, estimating a rotation period for the planet of 24.5 hours. Cassini's later observations refined this estimate to 24 hours, 40 minutes – incredibly close to the current-day value of 24 hours, 37 minutes and 23 seconds.

Giacomo Maraldi – Cassini's nephew! – studied Mars for over 40 years. He recorded a number of shape-shifting discoloured patches across the planet's disc, most notably a dark 'sea' and band that sliced across the planet (now known to be albedo features including the aforementioned Syrtis Major). He determined that that Mars's two polar caps were not perfectly aligned to the planet's rotational poles and appeared to change in shape and brightness over time, implying that the planet had seasons, and noticed a dark rim around the northern polar cap, which he attributed to 'meltwater' (actually a mix of windswept dust and layered deposits from when the caps seasonally change shape).

Between 1671 and 1673, Maraldi's Uncle Cassini worked with fellow astronomer Jean Richer to measure the distance to Mars. Richer headed to South America to observe the planet at opposition (aligned Sun–Earth–Mars in space), while Cassini echoed the observations from Paris. By viewing Mars from two separate locations on the surface of Earth, the astronomers were hoping to calculate this distance using an astronomical phenomenon known as parallax (the amount an object jumps when viewed along two different lines of sight, as seen when you peer at your outstretched finger through each eye independently). By observing from both South America and Paris, Cassini and Richer were increasing the distance between their 'eyes' so they could subsequently use geometry to figure out Mars's location. Together, the combined observations were used to calculate a distance to Mars that was just 7 per cent off the value we use today.

In 1698, Huygens's book *Cosmotheoros*, in which he took a philosophical stance on the possibility of life on other planets, was posthumously published (the astronomer passed away in 1695). He speaks of 'inhabitants' of Mars and listed many of the planet's properties, including an orbital period of 'about 687 days' and 'nights and days [that] have been found to be of about the same length with ours'. He also remarked on Mars's 'bigness': ''Tis remarkable that the bodies of the planets do not increase together with their distances from the Sun, but that Venus is much bigger than Mars' (this still puzzles astronomers – more on this later).

The 1700s brought a wealth of new knowledge about Mars and the 1800s were no different. One of the dominant astronomers in this time period was William Herschel (1738–1822). Herschel did a great deal for astronomy, most notably through his work developing the telescope, studying double stars, surveying large areas of the sky, and discovering numerous nebulae, multiple Solar System moons and the planet Uranus.

Herschel also fostered a curiosity for Mars, and often pondered the possibility of life in the Solar System. He turned his attention to Mars in the late 1700s. Referring to observations spanning from 1777 to 1783, Herschel remarked on the ever-shifting shape of Mars's polar caps, noting, as others had previously, that 'none of the bright spots on Mars [are] exactly at the poles, though they could certainly not be far from them'. He published a paper in March of 1784 in the world's first ever science journal, *Philosophical Transactions,* in which he explored Mars's polar caps, inclination, shape, size and atmosphere.

'The analogy between Mars and the Earth is, perhaps, by far the greatest in the whole Solar System,' he wrote. 'Their diurnal motion is nearly the same; the obliquity of their respective ecliptics, on which the seasons depend, not very different; of all the superior planets the distance of Mars from the Sun is by far the nearest alike to that of the Earth; nor will the length of the Martian year appear very different from that which we enjoy when compared to the surprising duration of the years of Jupiter, Saturn, and Uranus.'

In his paper, Herschel stated that Mars is an oblate spheroid – a sphere pressed slightly flat at the two poles – with a diameter of just over half that of Earth, an axial tilt of just under 30 degrees, a 'considerable but moderate atmosphere', and polar caps composed of ice, which partially melt when exposed to the Sun throughout the year and whose brightness is due to the 'vivid reflection of light from frozen regions'. He also (incorrectly) attributed various transient bright and dark belts and patches that he observed stretching across the face of Mars as being due to clouds and vapour in the planet's atmosphere, as some of his contemporaries had done previously.

While some of his observations and deductions weren't far from the truth – Mars's polar caps are indeed made of ice (although a mixture of water and carbon dioxide ice), the planet's axial tilt is around 25°, it is shaped as an oblate spheroid and its diameter is indeed just over half of Earth's – Herschel's ideas weren't always correct. For one, he was convinced that the Solar System was teeming with life, with every planet containing some kind of alien society. In fairness to Herschel, this was a prevalent view at the time and for many years after. However, Herschel took it a step further. He claimed to find considerable evidence for life on the Moon, documenting the existence of cities, roads, canals and vegetation, and even believed that the Sun hosted life, too. He described our star as 'nothing else than a very eminent, large and lucid planet … a cool, dark, solid globe clothed in luxurious vegetation and richly stored with inhabitants, protected by a heavy cloud-canopy from the intolerable glare of the upper luminous region'.

Herschel's determined belief that the Solar System planets were all inhabited by aliens fed the idea of life on Mars in the first real way. His observations of an unevenly shaded disc with dark and light patches led him to speculate on the idea of vast oceans of liquid water covering the planet, broken up by patches of rust-coloured land. In 1784, he wrote that Mars's inhabitants 'probably enjoy a situation in many respects similar to ours'.

Over the next 200 years, our knowledge of Mars became increasingly refined and its assumed similarity to Earth began to slip away. Various astronomers clarified what we knew about our red-hued neighbour – we spied yellow dust clouds and white hazy clouds, the shape-shifting polar caps were monitored and seen to grow and diminish with the seasons, the planet's rotational period was honed to its currently accepted value, numerous dark and light features on Mars's surface were mapped as 'continents' and 'oceans', and some of the first ever Martian crater maps were produced. Astronomers drew better and better maps of the planet's surface, showing bright and dark markings scattered across its disc in increasing detail. Many (correctly) suspected that the cratering seen on Mars's surface was due to past meteorite impacts. Astronomers found Mars's atmosphere to be significantly thinner than Earth's, and deduced that its surface temperatures and pressures must thus be far lower, too.

One of the key figures in the nineteenth century's exploration of Mars was Giovanni Schiaparelli (1835–1910), the dubious star of the previous chapter. During Mars's 1877 opposition, the Italian astronomer peered closely at the face of the planet, noticing his infamous criss-crossing system of 'canals'. He later went on to work closely with American astronomer Percival Lowell (1855–1916), one of the most fervent supporters of his *canali* hypothesis. Lowell studied Mars for years, establishing the Mars-focused Lowell Observatory in Arizona in 1894. He published several titles dedicated to the planet that excitedly detailed the many 'non-natural' features scattered across the planet's surface.

The scientific community at the time was divided and critical of the idea: some astronomers claimed to see the same lines, while others simply couldn't convince themselves of their truth. One of Lowell's colleagues, Carl Lampland, even purported to have taken a photograph of the Martian canals, but many of Lowell and Schiaparelli's contemporaries struggled to see any sign of them. Despite the scientific uncertainty, Lowell's work had a great impact on the public, and the idea survived well into the twentieth century. The astronomer spun an appealing narrative of a once-thriving civilisation that had desperately resorted to carving out shallow irrigation channels, in a bid to funnel water from the melting ice caps across the planet's parched surface.

In 1950, another theory was suggested for the formation of the canal system: Clyde Tombaugh, discoverer of Pluto, proposed that the canals might actually be the result of a web of fractures snaking across the planet. Instead of intentional canals, Tombaugh saw 'deep fissures of fractured land caused by asteroid impacts'.

As we now know, both were wrong – but the existence of Martian canals was only firmly disproved in the 1960s, when the US sent its Mariner Mars spacecraft on their historic photo-gathering quest. However, Schiaparelli's many hours spent gazing at and mapping the Red Planet had not been in vain. The astronomer named a number of observed features on Mars other than the *canali* and recorded them on a detailed map he completed in 1886. Schiaparelli christened what he thought to be seas, continents, large swathes of deserted land – which we now know to be simply different albedo features – and gave them names from Greek mythology. Seas and lakes became *mare* and *lacus,* land and plains became *planitia* and so on. Many of these names are still widely used today.

1877 was a fruitful year for Mars-loving astronomers. Alongside the birth of Schiaparelli and Lowell's canal craze, the Mars opposition of 1877 saw perhaps the most scientifically important Mars-related finding of the nineteenth century: the discovery of Mars's two small moons. American astronomer Asaph Hall observed the two moons from Washington DC in August of 1877 – he spotted the smaller and outermost moon as a 'faint object' on 11 August, before being forced to stop observing by cloudy weather. He waited impatiently for clear skies. 'On August 16 the object was found again on the following side of the planet ... it was moving with the planet,' he wrote. 'On August 17, while waiting and watching for the outer satellite, I discovered the inner one.' Hall went on to name the inner and outer moons Phobos and Deimos, respectively.

Another prevailing opinion in the 1950s and 60s was that of Martian vegetation. Various astronomers – Tombaugh, Lowell and his colleague William Sinton, even popular science fiction author H. G. Wells, as described in a previous chapter – threw their voices behind the idea that Mars was covered in leafy green plants, which bloomed and thrived during the spring and summer months, and faded in the winter. A wave of darkening was observed to sweep across the face of Mars every summer, thought to be due to vegetation and/or bodies of water, and the polar caps appeared to seasonally melt, perhaps providing much-needed water to any plant life in the area. While some attributed the colour changes to changing cloud cover, as Herschel did, some envisaged entire jungles of towering trees that expanded across Mars's surface during the summer. 'It appears very likely that there is organic matter of some kind present on the surface of Mars,' wrote Sinton in 1958. Tombaugh agreed but was less optimistic about the prevalence of sizeable plants, instead claiming them to be lichens due to the 'extreme heat, cold, and lack of water'.

We now know that this 'wave of darkening' is actually due to albedo features – patches of Mars that reflect incident sunlight by different amounts – changing in colour over time. These colour changes may be driven by dust rippling and blowing across the planet in seasonal wind patterns, coating some areas with lighter material and unearthing darker patches. Albedo can also change seasonally due to factors such as illumination angle, soil composition, space weathering and moisture. The polar caps do indeed disappear during the Martian summer and accumulate during winter; in the colder months a metre-thick layer of dust and frozen carbon dioxide covers the brighter residual caps, making them appear larger.

While we had formed a pretty good picture of Mars in the 1950s, based on centuries of stargazing and scientific study, some myths still persisted. This speculation was swept away by the launch of NASA's Mariner Mars missions in the 1960s. These missions, combined with data from the later Viking probes in the 1970s and beyond, gave us the first glimpse at the reality of Mars, and put a firm end to the idea of canals, liquid water and advanced or intelligent life.

Among other things, these missions found a Mars that was uninhabited, arid and dry. It had a thin atmosphere composed largely of carbon dioxide (confirming earlier observations in the 1940s that had suggested a carbon dioxide-rich atmosphere). The pressure on the surface was incredibly low, measuring less than 1 per cent of Earth's. *Mariner 4* detected freezing cold temperatures – around -100°C (-148°F) in daytime – but the later missions discovered that the planet could also be surprisingly temperate, albeit changeable (the currently-accepted temperature ranges between around -150 and 30°C (-238 and 86°F) depending on location and time of day, with an average of around -60°C (-76°F)).

Our probes couldn't find any sign of a magnetic field, implying that the Martian surface was a very inhospitable place to be (we believe this to be true: that Mars's intrinsic magnetic field is no longer present, but likely was in the past). Without a magnetic field, Mars would have no protection against radiation from space.

This view was to form the basis of our understanding of Mars for decades. Subsequent missions refined our knowledge more and more, exploring everything from the composition and structure of the soil (slightly alkaline and rich in iron oxide, causing the planet's characteristic colour) and surface rocks (primarily basaltic, both igneous and sedimentary), surface features (everything from large craters to sand dunes to towering volcanoes), seasonal variations (albedo changes, polar melt, seeping brines locked up within the soil), geology (once-intense volcanism and lava flows, avalanches, possible plate tectonics), water (once present in huge amounts, but now locked up in the polar caps, as permafrost and briny flows within the soil, perhaps deep underground), internal structure (an iron-rich core, perhaps a mix of molten and solid material, wrapped in a thick crust), diameter (half [0.53] that of Earth), surface gravity (just 38 per cent of Earth's), crater history (extensive!), atmosphere, orbit, axial tilt (25.2°, slightly more than Earth), climate – anything and everything you can think of.

We now know more about Mars than we've ever known, but there's still much left to explore – and the planet still has a number of mysteries up its sleeve.

CHAPTER SIX

The Massive Mars Problem

The Mars story began some 4.6 billion years ago, when the Universe was two-thirds of its current age. Instead of a family of planets encircling a sunny star, a giant cloud filled our entire region of space and far beyond: a cold, lumpy, dense clump of gas and dust gently turning in the emptiness of space.

Over time, this unstable cloud began to collapse inwards due to its own gravity; it's possible that a particularly unstable part of the cloud randomly collapsed, or that the violent death of a star near to the Solar System sent a shockwave hurtling in our direction and caused an instability to begin crumpling in on itself. The cloud contracted, shrinking and flattening, spinning faster, becoming hotter and denser. Much of the mass within the cloud flowed inwards towards the cloud's centre, building up until it became hot and dense enough to spark the formation of a new star: the infant Sun!

This Sun was very different to the star we know and love today. Rather than a mature adult, the young Sun was unstable and rebellious. The Sun pulled its basic form together in under 10 million years, but spent a further 50 million cycling through the various astronomical hurdles on the road to adulthood. It went from protostar – the very first stage of stellar formation, when a baby star is still hungrily feeding from its surroundings – to something known as a T Tauri star – a rebellious and variable state in which the star throws off powerful jets of intense radiation and thick winds – to pre-main-sequence star – where the star is pretty close to its final mass but hasn't yet ignited nuclear fusion – and, finally, main-sequence star, where it remains today. Now powered by the fusion of hydrogen into helium in its core, the Sun is roughly halfway through its 'adult' life, and has about

5 billion years left to go before the next (and final) stage of its life begins.

In its earliest days, the Sun was a protostar enveloped by a thick blanket of dust and gas. While a significant portion of the original cloud's mass – well over 95 per cent – was consumed by the hungry newborn star, at least a few per cent of the mass remained in the form of a protoplanetary disc. This disc was a substantial saucer of hydrogen, helium, ice and silicates that encircled the Sun's equator and stretched out for tens of billions of kilometres. The Sun continued to feed from the inner edge of the disc throughout its adolescence, pulling and dragging at the gas and dust there with its gravity and causing material to spiral down on to its surface.

This disc likely had a similar composition to the Sun itself, with roughly 99 per cent of the mass in the form of gas (primarily hydrogen and helium) and the remaining per cent as solid grains of dust. These dust grains were ten times thinner than a human hair at their smallest, on the scale of a thousandth of a millimetre (a micrometre, μm). However, these tiny bacterium-sized dust grains were by no means insignificant. They were to become the building blocks for all of the planets, asteroids and moons in the Solar System.

The classic idea for terrestrial planet formation – that is, for the rocky planets sitting within the asteroid belt – is remarkably similar to a giant game of cosmic Lego. Small dust grains whirled around within the disc, colliding with one another and clumping together. Over time they grew larger and larger until they formed kilometre-sized asteroid-like bodies known as planetesimals.

These planetesimals began to gravitationally interact with one another, colliding and sticking, or smashing together, shattering and sweeping up the subsequent debris. Bigger bodies began to grow faster than their smaller relatives, scooping up smaller planetesimals more quickly, and eventually reached sizes such that a single large body completely dominated their region of the disc (stages known as runaway and oligarchic growth respectively).

These growth processes continued until most of the embryos were roughly the size of Mars. Then began another battle! These large embryos smashed into one another more violently than ever before in giant impacts, with the winner subsuming their defeated rival to grow even larger still. In the outer Solar System most of the material was icy and rotating relatively slowly, meaning that the planetary embryos collided a little less violently and could stick together more easily, but in the inner Solar System the material was both drier and moving faster due to its proximity to the Sun. As a result these collisions were more destructive: objects collided and shattered one another into pieces, before accreting the debris. Some smaller planetesimals were not destroyed but merely banished, flung out of the system by a larger rival's gravity (a process known as 'dynamical clearing'). At the end of this cosmic war, just four victors remained in the inner Solar System. These lucky embryos later formed Mercury, Venus, Earth and Mars.

Bodies in the outer Solar System underwent a similar process until they became massive enough to also pilfer gas from the surrounding disc. They wrapped this gas around themselves, eventually becoming the gas giant planets. The protoplanetary disc would have dispersed completely within a short timeframe, meaning that the giants must have formed and cloaked themselves in gas in just a few million years or so.

Although there are still many unknowns, this is the standard picture of how our Solar System formed and it has been widely accepted for years. However, it has several fundamental problems, one of which is Mars. 'The origin of Mars is a fascinating subject, but also a frustrating one,' says John Chambers of the Carnegie Institution for Science in Washington, US. 'I think it's fair to say that there's no consensus in the scientific community about precisely how it formed, and why it has some key differences from Earth.'

Mars is a puzzle in a number of ways, two in particular. Firstly, Mars appears to have formed really quickly, in just a few million years, but we know that it took Earth roughly

100 million years. Secondly, if you look at the planets in the inner Solar System, they follow one pretty obvious pattern: they increase in both size and mass as their distance from the Sun increases. Venus is larger than Mercury, Earth is larger than Venus, and Mars … well, Mars is significantly smaller and less massive than both Earth and Venus. In fact, the Red Planet only holds 10 per cent of the mass of Earth, even though our models show that it should hold the same amount or more. 'It's possible that Mars – and Mercury? – represent embryos that never underwent giant impacts, so their growth stopped earlier than Earth and Venus,' adds Chambers, 'although we don't know why.'

Mars's size goes against everything we thought we knew about terrestrial planet formation. This is known as the 'massive Mars problem', and has frustrated scientists for decades (and is still doing so!). It's also acted as one of the catalysts for a recent and sizeable shift in how we perceive the early Solar System. 'Our fundamental picture of planet formation is undergoing a significant change at the moment,' explains Chambers. 'Several scenarios have been proposed to explain why Mars is smaller than Earth and Venus, but none is widely accepted yet. Researchers keep coming up with new ideas, too. None of our models are particularly well developed at this point, so it's hard to know which may be correct.'

There are a few hypotheses worth considering, some of which require more precise timings and happy coincidences than others. In general, our theories for a small Mars fall into two categories: either something somehow removed the majority of the Lego blocks from the region where Mars (and the asteroid belt) currently exists, placing an upper limit on Mars's size, or something stopped Mars from growing efficiently even though there was plenty of material for the taking. To add another layer of complexity, any theory that can successfully recreate the size and mass of Mars must also be able to recreate the architecture of the entire inner Solar System, including the asteroid belt.

Currently, there is no leading theory for how Mars formed: there's no single idea that the majority of scientists throw their voices behind, as the entire time period is currently so unclear. 'This is a particularly hectic period, in which people are hastily throwing around new ideas,' agrees Gennaro D'Angelo of the SETI Institute in California, US. Figuring out how Mars formed is 'no easy task'!

First up is the classical model, the hit-and-run scenario, which plays out exactly as described above (and as the name suggests). In this model, Mars gathered up all the material in its vicinity and grew larger over time via ongoing hit-and-run collisions, smashing other bodies to bits and accreting the resulting fragments.

Among other assumptions (including the number of embryos, number of planetesimals, their distribution and the size of the overall disc), this theory models the protoplanetary disc in such a way that the disc contained more and more mass with increasing radius. (This is one of the reasons the gas giants could grow so much more massive than, say, tiny Mercury.)

However, this model does nothing to solve the problem of Mars's mass. It produces a large, bulky Mars; there's no easy way for it not to, as Mars's environment simply contained more stuff for Mars to consume than the environments of Mercury, Venus and Earth. If there was more mass in an outer region than an inner one, there should be more mass in the region around Mars than around Earth, and Mars should be the more massive of the two. 'And that's exactly what comes out of our simulations,' says Nader Haghighipour of the University of Hawaii. 'In all simulations, the mass of Mars is larger than that of Earth. However, while this agrees with physics and makes perfect sense, it doesn't agree with reality.'

Although we're mainly focusing on how the terrestrial planets formed, this model is also problematic for forming the gas giant planets. We've observed protoplanetary discs around other young stars and seen that they only seem to survive for a few million years. While this is sufficiently long to form the basis of a planetary embryo, it isn't long enough for bodies to grow massive enough to snag and accrete gas – this requires a mass of around 10 times the mass of Earth at least. Other models have tentatively been suggested to fix this problem (pebble accretion, described later, being one).

However, this model is appealing because it requires almost no fine-tuning, no one-of-a-kind events or serendipitous timings, and is based on solid physics. Despite its problems, it could potentially also form a small Mars; although incredibly unlikely, it's possible that Mars is just an unexpected and very low-probability outcome from such a model. If we simulated the starting conditions for our Solar System and ran the scenario many, many more times – tens of thousands of times – a system that looks like ours might end up popping out due to pure luck (bad luck, in Mars's case).

Our observations of other planetary systems do suggest that this may be the case. We've discovered thousands of exoplanets – as of August 2016, over 3,500 – orbiting around thousands of different stars, but we've not found a single system that looks like our own. This could be due to some other as-yet-unknown cause, but is more likely to indicate that our Solar System, just like Mars, is an oddball.

To tackle the problem of the giant planets – namely that they can't grow massive enough to trap gas within a protoplanetary disc's lifetime – some have suggested changing our view of the building blocks themselves. This is the so-called pebble accretion model. Rather than having tiny dusty grains whirling round within the disc and sticking together to grow ever-larger, pebble accretion proposes that a large amount of mass already existed within the initial disc in the form of pebble-sized particles. 'There is some observational evidence to suggest this is true in other protoplanetary discs,' says Chambers.

Unlike planetesimals, pebble-sized particles would be small enough to be strongly affected by the gas within the disc, something that would be particularly important when they happened to closely approach a large planetary embryo or planetesimal. During a close encounter, gas would drag at the pebble and slow it down significantly, greatly increasing the chances of it being pulled into the larger body and speeding up the forming planet's overall growth. 'It seems likely that this is how the giant planets in the Solar System formed, which is quite different than the standard model,' says Chambers. 'However, it's still unclear whether pebble accretion played an important role in the formation of the terrestrial planets, and if this had something to do with the small size of Mars.'

While this idea has promise it's far from being fully formed and key questions remain. For example, it assumes the biggest pebbles to be a few metres in diameter, which are more like boulders or cobbles than what we would consider pebbles. How did these form in the first place? Additionally, these pebbles must have formed at a specific rate and within a specific timeframe; while they had to form quickly enough that the gas giants could bulk up rapidly, they couldn't have formed too quickly, as the forming planetary cores still needed time to interact with one another and battle with their competitors. If they formed too fast we'd have an outer Solar System flooded with icy, Earth-sized bodies, not a few mammoth ones.

Instead of the issue being the composition or size of the early building blocks, the problem may actually lie with their distribution – or so says the depleted-disc model. Rather than modelling the initial disc as a smoothly distributed ring of mass that increases with distance, the depleted-disc model instead assumes that the mass within the disc was unevenly distributed from the very beginning. Rather than being smooth and homogeneous, the disc was bumpy, with some patches naturally containing more mass than others; a combination of crowded and sparse areas all mixed together.

In this model, planetary embryos formed within a bumpy disc and then took up to three million years to self-adjust and settle into a distribution similar to that modelled in the standard case, one that was uniform and had more mass as distance increased. This self-adjustment would have happened as the planetesimals interacted with one another in the early Solar System, causing many bodies to have likely been flung into Mars's vicinity from different orbits, filling the gap and smoothing out the disc's mass distribution over time. Mars would have formed quickly during this 'settling' period.

'In such a disc, Mars formed in one of the regions where the mass accumulation was higher than its surroundings,' explains Haghighipour. 'As soon as Mars interacted with other bodies in that region, it was scattered to a neighbouring region with less mass, and then stopped growing because there wasn't enough mass in its orbit to accrete.'

This model has its roots in a 2009 study from UCLA in California, US, which explored what would happen within a smaller, constrained disc: a dense ring of planetary embryos that sat just between 0.7 and 1 AU, rather than stretching out continuously through space.

In these simulations, the embryos within the ring grew and interacted with one another as expected, altering their orbits and moving around. In some cases, Mars ended up getting shunted out of the material-rich ring into a sparser region just outside of it, where there was nothing to accrete. Mars then maintained its mass, orbit and size as there were no other rivals to disrupt it, leading to a result somewhat similar to what we see in reality. Although the model was unrealistic in that it only considered a small portion of the Solar System, it showed that, in theory, a planet like Mars could form if we changed the conditions of the initial disc itself. 'A local depletion is relatively easy to justify, since there's no indication that the initial distribution of solids in the region of the terrestrial planets had to vary smoothly with distance from the Sun,' says D'Angelo.

While this model manages to simulate a small, impatient Mars that completed its growth and formation promptly, it is not fully developed for the entire Solar System, leaving it currently very incomplete. However, it does not require a great deal of tweaking or assumption, making it a promising, and relatively intuitive, option.

Lastly comes the grand tack hypothesis, which made a big splash in 2011. As with the depleted-disc model, the grand tack supposes that Mars formed in a mass-rich region of the disc and later ended up in a sparser region, which abruptly stopped its growth. However, there are a few key differences – the main one being the cause of the bumpy, irregular mass distribution through the disc.

While the former models the disc as being naturally bumpy and inhomogeneous from the beginning, which then affects the planets it spawns, the grand tack has a more exciting proposition: that the disc was completely disrupted and altered when Jupiter careened inwards, blasting into and completely reshaping the inner Solar System!

This hypothesis likens Jupiter to a yacht 'tacking' through space, hence the name. It proposes that, after forming at a distance of around 3.5 AU from the Sun (closer than its current orbit of 5.2 AU), Jupiter came hurtling inwards towards the Sun, speeding towards our star and affecting everything in its path. Saturn later did the same, moving inwards after its larger relative. The two snowploughed inwards, pushing and shoving the material in their way and scooping it up with them as they went, causing it to stockpile closer to the Sun than it had been previously.

Once Jupiter reached 1.5 AU, just beyond the orbit of Earth, Saturn's influence took hold of Jupiter*, forcing the gas giant to stop and turn around. The duo then sped right back outwards to settle in their current orbits. This dramatic migration truncated the inner Solar System's planet-forming disc to within the present-day location of Earth (1 AU), creating a scenario similar to that in the 2009 depleted-disc model.

This solves the massive Mars problem. It severely limits the amount of mass available for Mars to accrete, naturally

* This was possible due to a dynamical relationship known as an 'orbital resonance' – essentially, the two planets orbited the Sun in such a way that their orbital periods had a specific ratio, allowing them to influence one another gravitationally. Other bodies in the Solar System have such resonances today, notably Pluto and Neptune (2:3, meaning that Pluto completes two orbits for every three of Neptune's), and Jupiter's moons Ganymede, Europa and Io (a 1:2:4 resonance).

stunting its growth and essentially freezing it in time as a planetary embryo.

As Jupiter blazes both inwards and back outwards through the asteroid belt, disrupting it twice, it flings bodies in the same directions, both inwards and outwards. This influence would result in the asteroid belt comprising a scattered population of bodies that formed both within and beyond the present-day location of the belt, travelling on orbits with various inclinations and eccentricities, which is indeed what we see. It'd also remove a great amount of the mass from the region, explaining why there's no planet there.

As another bonus, the grand tack could explain why the Solar System looks so different to other systems we see. Many exoplanetary systems have huge planets orbiting incredibly close to their host star, often within the orbit of Mercury! We don't see this, but perhaps Jupiter is to blame. When it migrated inwards it would have caused chaos, forcing any larger bodies to collide and fracture and to be eventually dragged inwards to collide with the Sun, disappearing without a trace.

Despite its success in many areas, there are problems with the grand tack that need clarification, mostly concerning the fact that the hypothesis requires a lot of things to come together at precisely the right time, and a number of levers to be pulled exactly on cue. 'The grand tack model is neither natural nor intuitive,' comments Haghighipour. 'It requires major fine-tuning and very specific assumptions. If those conditions and assumptions aren't there, the scenario doesn't work.'

However, the grand tack may be a correct interpretation of our Solar System – and, as Chambers points out, that the hypothesis requires 'exquisite timing' and numerous caveats 'doesn't necessarily mean the model is wrong, but it would predict that systems of terrestrial planets like our own are rare elsewhere'.

Other models have also relied on Jupiter as a way of removing mass from the inner Solar System. For example, one possibility is that Jupiter once had a far more eccentric orbit than it does today, meaning that it'd approach the Sun far more closely. During these approaches the giant planet might have scooped up or scattered many of the planetesimals from Mars's habitat, accreting some and throwing others out of the disc entirely. But, unsurprisingly, there are problems with this idea: if Jupiter affected the region around Mars so strongly, it would have also affected the asteroid belt, which sits between Mars and Jupiter, and cut down the amount of water-rich objects residing there. As a result, few watery bodies would have existed in the region around Earth (and Mars). If this theory for the formation of Mars were true, we would struggle to explain why both our planet – and early Mars – were so wet.

The three faces of Mars

Maybe it was a pebbly beginning, maybe a rogue Jupiter, maybe a straightforward accumulation of cosmic dust. However it formed, at this point in the story we have an infant Mars, ready and raring to begin its young life.

Early Mars was a very different beast to the Red Planet we see today. However dry it may be now, we believe that it was once a very wet and watery world*. We see surface features – valleys, channels, gullies, riverbeds, even potential seabeds and coastlines – and minerals that could only have formed in the presence of water. Early Mars's surface was likely covered in streams, lakes, rivers, maybe even an ocean. This is one of the biggest unsolved problems in Martian geology (or areology!): how could conditions have existed on early Mars to support such a huge and widely distributed amount of water, and where on Earth – or, rather, Mars – did it all go?

Liquid water can't survive on Mars today as the surface pressures and temperatures are simply too low (and the latter too variable). Water cannot exist on Mars as a liquid, instead

* We believe this water to have been delivered by asteroids and comets during Mars's earliest days (hence why theories that remove water-rich material from Mars's vicinity are difficult to reconcile).

quickly evaporating, freezing or sublimating (changing straight from a solid to a gas). However, in its early days, both Mars's surface and atmosphere were very different to what we see today.

Mars's geological history is usually split into three main time periods: the Noachian, Hesperian and Amazonian. Each period is named after a patch of the Martian surface we believe to have formed during that time. It's difficult to determine exact ages for these periods, but we have some best estimates based on the features we see. Unlike Earth, whose surface has been continually changed and resurfaced and evidence from past eras essentially smoothed over and hidden, large parts of Mars's surface remain relatively ancient and preserved. Much of the surface is at least 3.5 billion years old!

The Noachian period began some 4.5 billion years ago with the formation of Mars, and stretched through a period of time dubbed the Late Heavy Bombardment (anything that occurred before roughly 4.1 billion years ago is sometimes known as 'pre-Noachian'). During this event, the rocky planets in the inner Solar System were peppered and bombarded with an unexpectedly high flux of rocky impactors. These may have been swept up and pushed inwards by the gas giants, resulting in a flood of rocky bullets headed towards Earth, Mars and neighbours. The terrestrial planets ended up with uncountable impact craters*, many caused by impact events that were destructive and energetic enough to literally melt rock ('impact melt'), and numerous large impact basins. This period of time is named after Mars's Noachis region (Noachis Terra), a particularly heavily cratered and rugged part of the Red Planet's southern hemisphere that formed in the Noachian period.

Noachian Mars was very different to the planet we see today. It saw intense volcanic activity and was warmer, wetter

---

* The largest craters on Mars are the Borealis (8,500km or 5,300 miles in diameter) and Utopia (3,300km or 2,050 miles in diameter) basins in the northern hemisphere, and the Hellas basin (2,300km or 1,420 miles in diameter) in the southern hemisphere.

and more humid; we see signs of water erosion, channels, deltas, valleys, shorelines, large lake basins and potential riverbeds. It likely rained and snowed! Present-day Mars is cloaked in a thin, tenuous blanket of mostly carbon dioxide. To support huge ancient bodies of liquid water, including a proposed sea that covered up to a third of the entire planet, and an ongoing precipitation cycle to replenish them, this atmosphere must have been significantly thicker than we see today. Compounding this issue is the fact that the early Sun was around 30 per cent less luminous than it is now, requiring an atmosphere that was thicker still in order to sufficiently raise the temperature. Such an atmosphere would have raised the surface pressures and temperatures significantly, allowing the planet to cling on to its water.

Alongside physical signs of past water in the form of valleys and branching river deltas, we also see indications in Mars's chemistry. The presence of water can change certain substances considerably and facilitate the creation of others that are just unable to form on a dry, arid world.

For example, ESA's *Mars Express* and NASA's *Mars Reconnaissance Orbiter* (MRO) have both explored various patches of Mars's surface (including pre-Noachian and Noachian rocky outcrops) and detected hydrated sulphates, phyllosilicates – rock-forming minerals based on silicon and oxygen, found in abundance in Earth's crust – and clay minerals (containing aluminium, iron, magnesium, olivine), all of which need water to form. NASA's *Curiosity* and *Opportunity* rovers have also found various pieces of evidence for past water on Mars, including specific rock chemistries, textures and mineralogies that could only have been created by coming into contact with water.

We have also discovered seasonal briny deposits streaking down the walls of Martian craters, known as 'recurring slope lineae' (RSL) – in 2015, MRO analysed the RSLs and detected the signatures of hydrated minerals, confirming that water was key in forming the features. Rather than talking of ancient water on Mars, we are now able to talk about water, albeit salty, actually flowing on the planet's surface.

Overall, NASA has concluded that the planet was once 'soaking wet' and 'drenched in liquid water', and that it still has some stores of water in some form or other today (as permafrost, ice or salty brines laced through or just below the Martian soil, in the polar caps and maybe underground).

After this initial injection of water in the first billion years or so, Mars's surface became far drier and more arid and only experienced wet conditions sporadically, with catastrophic flooding events sweeping across the entirety of the planet. Noachian Mars may have formed underground hydrothermal vents and hot springs – warmed by volcanic processes also occurring at the time – which fed warm pockets of subsurface water.

Recent studies (2016) have even suggested that colossal waves of water sloshed around on ancient Mars's surface, building into immense tsunamis that put Earth's to shame! These tsunamis may have been triggered by particularly violent meteorite impacts, resulting in at least two separate occasions where 45 to 120m (150 to 400ft) waves surged across the planet, reshaping its geology, moving material around, and changing Mars's apparent coastlines. However, these 'coastlines' are still unconfirmed. Astronomers propose that Mars may have once had a huge ocean (dubbed Mars's Paleo-Ocean, or Oceanus Borealis) sitting in its northern hemisphere during the Noachian period, filling a region known as Vastitas Borealis (literally 'northern waste'). Vastitas Borealis covers much of Mars's northern hemisphere. Its terrain dips surprisingly low into Mars's crust and is far flatter and smoother than its surroundings: characteristics that are reminiscent of a seabed. We see what we believe could be coastlines and eroded cliffs, gullies and channels by which such a sea could have been replenished.

This is still a controversial idea: while some of our observations support the hypothesis of a Paleo-Ocean, others do not, or remain ambiguous. Such a large ocean may have sublimed and been lost to the cosmos, may have seeped down through the soil and retreated to the subsurface, or may have frozen over and been covered by sediment and soil. If either of the latter two fates were true, a huge reserve of water may still exist underneath the surface of Mars's northern hemisphere. To resolve the debate we'll need to send more probes to Mars and see what we find.

The two key present-day water stores on Mars sit at the planet's poles. Mars's poles are covered by permanent blankets of white ice – a mixture of water and carbon dioxide ice – which partially thaw in the summer and refreeze in the winter. The northern cap is mainly water ice with a thinner metre-thick covering of frozen carbon dioxide (dry ice) that only appears in the winter, while the southern cap comprises water ice covered in a permanent and far thicker dry-ice layer (many metres thick). The northern cap is bigger – 1,000 versus 350km in diameter (621 versus 217 miles) – but Mars's south polar region alone is thought to contain so much water ice that, if melted, it could flood the entire planet to a depth of 11m (36ft)!

These polar ice reserves may not have always been in their current locations. Mars's poles are thought to have shifted to their current position in the Noachian era. Early geologic processes – namely the massive and dominating volcanism of the Tharsis dome, a feature now situated on the Martian equator – may have caused the crust of the planet to swivel on top of its core by up to 25° (wonderfully described by the associated 2016 press release as 'like turning the flesh of an apricot around its stone'). While the tilt of the planet itself didn't change, the surface features would have completely realigned themselves. This may partially explain why some suspected buried water reserves (subsurface ice, glacial melt) on the planet appear to be located far from the poles, and could explain why water seems to have flowed in otherwise unusual places on the surface.

While the idea of a warm and wet Mars that slowly became arid, as described here, is generally accepted as both possible and likely, there is also a growing view among some scientists that Mars only experienced episodic warm and wet conditions – bursts that may have been driven by early

Martian volcanism. This is currently a less clear and less comprehensive picture of Mars's history, but it may well be true (we need to know more about how Mars has evolved in order to reach a conclusion).

If so, it means no less exciting things for early Mars and for life; episodes of warmth and moisture, fluctuating with the intensity and prevalence of spluttering Martian volcanism across many millions and billions of years, could have created pockets of habitability across Mars. Much of the water-related phenomena described here – ongoing erosion, hydrothermal vents and hot springs, changing polar caps, ancient riverbeds, potential oceans and shorelines – could also fit into this picture. For a planet to be habitable or develop life, it need not be continuously habitable across time; isolated bubbles of warmth and water – in other words, life-friendly conditions – may have existed across Mars at varying points in the planet's history, even if they were changeable or short-lived (more on potential Martian life in a later chapter).

Specifics aside, we know that the climate of Noachian Mars was warm enough to support plentiful liquid water – but that quickly changed. Towards the end of the Noachian era, something fateful happened that reshaped the entire planet: Mars lost its atmosphere.

ESA's *Mars Express* and NASA's *MAVEN* (Mars Atmosphere and Volatile EvolutioN) and *Curiosity* missions (among others) have investigated the curious case of Mars's missing atmosphere. Where did it all go? We know that water hung around on Mars for the first billion years of Mars's life, so the planet's atmosphere – and thus its magnetic field, as explained later – must also have been present for much of that time in order to keep everything stable.

The atmosphere could primarily have disappeared in one of two ways: up into space (to the atmosphere or wider cosmos), or down into the ground (to become locked up within rocks and soil).

MAVEN's observations suggested that 'escape to space', where the solar wind ionises and strips away material from the upper parts of a planetary atmosphere, was probably to blame. This process is still taking place today with what's left of the Martian atmosphere.

The solar wind streams into Mars's vicinity, interacts with its atmosphere and removes its uppermost ions, dragging them out into space*. The early Sun was also far more volatile and active, flinging out bursts of radiation and cultivating solar storms far more frequently. This could have made any atmospheric stripping up to 20 times more devastating! Once Mars's magnetic field switched off, after half a billion years or so, the planet's upper atmosphere would have been naked and vulnerable, lacking a cosmic shield. The solar wind could have streamed in and ionised molecules at a far higher rate, stripping them away faster.

This is a key difference between Mars and Earth: while the two terrestrial planets seem superficially similar, Earth has a magnetic field that helps us to protect and retain our atmosphere, whereas Mars lost that advantage a long time ago.

There's also the fact that Mars is smaller, less massive and thus has lower surface gravity than Earth. As a result, the planet simply struggles to keep hold of as much atmospheric material as we can. Mars's atmosphere may have also been blasted out into space by the destructive impacts we know to have taken place within the Noachian period (which in turn would have made subsequent impacts even more destructive due to the thinner Martian atmosphere, continuing and worsening the cycle). In a less exciting scenario, Mars's atmosphere might have just bled off into space of its own accord – the molecules may have swirled around and around, colliding and transferring energy between themselves, growing faster and faster until they reached escape velocity (this is a very crude explanation of a complex dynamical process known as 'Jeans escape').

---

* Data from ESA's *Mars Express* have indicated that escape to space might not have been the only culprit, as the stripping rates appear to be low and may not have been significantly different in the past. It's possible that Mars's atmosphere was stripped by a different process, or a mixture of several.

At least some of Mars's atmosphere is thought to have headed down into the planet's crust, interacting with the soil and binding to the elements present there (a process known as 'sequestration'). We see signs of this process in the few Martian meteorites we've found here on Earth, and NASA's *Curiosity* rover saw further signs when it arrived and began to study the soil and atmosphere in earnest. However, we have found that this is unlikely to have been a dominant process; we'd expect the heavier parts of Mars's atmosphere to have disappeared if it were, as they would have sunk down and been the ones to interact with the Martian surface. We see the opposite: instead of an atmosphere filled with lighter isotopes of the constituent elements, we see heavier ones (for example, carbon-13 rather than carbon-12, with the numbers representing the number of neutrons present).

This suggests that Mars's atmosphere was stripped from above and not below, thus removing more of the lighter elements and isotopes and leaving the heavier ones behind. We'd also expect to find carbonates, compounds that are a consequence of sequestration, in the Martian soil, but despite our various searches these don't seem to be present in large amounts.

We're still unsure of exactly what happened with Mars's atmosphere, but are pretty confident that the Red Planet was once wrapped in a far thicker atmosphere that allowed the planet to stay warm and pressurised enough to support bodies of running and standing water. Mars's ancient atmosphere was likely different in composition, with a far higher oxygen content (today it has just 0.13 per cent – its atmosphere comprises 95 to 96 per cent carbon dioxide, a couple of per cent each of argon and nitrogen, and trace others). We know this because we've found minerals in old Martian rock (manganese oxides) that need an oxygen-rich environment, or one with microbes present, to form (and we don't believe it to be the latter).

Mars lost the majority of its atmosphere when it was less than a billion years old, and embarked upon the next phase of its life: the Hesperian period. This period dates back to between 3 and 3.7 billion years ago and is named after

Hesperia Planum, a prominent lava plain located in Mars's southern hemisphere. Just as the Noachian years were famous for the sheer amount of cratering and abundance of water they experienced, this period of time is characterised by its volcanism!

Mars transitioned from being watery to being predominantly dry at some point within the Hesperian period, and the planet's atmosphere thinned to the level we see today. The barrage of cosmic impactors finally wound down and volcanic activity took over in shaping the surface of the planet*. In fact, NASA's *Curiosity* rover recently found traces of a mineral (tridymite) that's generally thought to only form through particularly hot and explosive forms of volcanism (silicic volcanism), so it's possible that Mars's volcanic history may be even more intense than we think.

Mars is famous for having some of the largest volcanoes in the entire Solar System, many of which formed within the Hesperian period or earlier. Mars's volcanism is now confined to a couple of key regions on its surface – Tharsis and Elysium – with a few other volcanic features dotted around.

The Tharsis region is massive. It spans Mars's equator and covers around a quarter of the planet's entire surface area. The elevated plateau houses some of the largest volcanoes in the Solar System (including three colossal volcanoes named the Tharsis Montes – Ascraeus Mons, Pavonis Mons and Arsia Mons – and the famous Olympus Mons, which lies just to the north-west of the plateau). The Tharsis bulge is thought to sit atop a volcanic-tectonic hot spot like that beneath Hawaii. The other key patch of volcanism is Elysium, a region in Mars's northern hemisphere that contains a clump of three other volcanoes (Hecates Tholus, Elysium Mons and Albor Tholus).

---

*The Noachian period also saw intense volcanism, but the Hesperian really saw the effects of volcanism take hold of Mars (volcanic plains formed, the planet experienced vast resurfacing, new atmospheric and rock chemistries emerged and more).

Olympus Mons is the largest volcano discovered on any planet in the entire Solar System, standing at around 25km (15.5 miles) tall. Its footprint would cover all of the American state of Arizona, or a large portion of France. For reference, the very deepest part of Earth's oceans is the Mariana Trench with a maximum depth of just under 11km (6.8 miles) and Mount Everest's peak is 8.8km (5.5 miles) above sea level. Even Earth's most extreme features are tiny in comparison!

Martian volcanoes released immense flows of lava that covered large portions of Mars's surface and smoothed it out, creating vast plains, ridged terrain and lava tubes (which are exactly what you might expect: tubes that drain and transport lava beneath a planet's surface). Hesperia Planum is a good example of a lava plain scattered with features named 'wrinkle ridges'. These ridges, reminiscent of veins or stretch marks, form on basalt-rich lava plains either when the lava itself begins to cool down, contracting as it does so, or when tectonic plates push and fold over or against one another.

Martian volcanoes may be able to grow so unimaginably large due to the planet's likely lack of plate tectonics, which break up and disrupt features as they grow. We're still unsure, and predominantly sceptical, about the presence of plate tectonics on Mars: namely whether or not the planet has numerous plates like the shattered eggshell of Earth, or if its crust is in one large intact piece. We know that Mars lacks tectonics today, but there have been various signs suggesting that the planet may have had them at one point in its history, perhaps in the mid-late Noachian period or early-mid Hesperian. On Earth, tectonic plates shift around, rub against one another, overlap, dig beneath other plates, and create long chains of mountain ridges, volcanoes, ocean trenches, valleys, chasms and fault lines. It's possible that some of these processes once took place on Mars, until rapid cooling froze the process at a very primitive stage of development.

Various processes and features suggest this might be the case, including the huge canyon system discovered by NASA's *Mariner 9* probe, Valleris Marineris. This is a huge,

4,000km-long, 200km-wide, 7km-deep system of valleys and canyons marking Mars's equator east of the Tharsis region (2,500 miles long, 125 miles wide, 4 miles deep). It is strikingly obvious in any photograph of Mars. Its creation is thought to be related to the Tharsis region – Mars's crust may have cracked and split apart as Tharsis formed – or to early plate tectonics followed by ongoing water and lava erosion. We do see evidence of water erosion in some parts of Valles Marineris, making it an especially interesting location.

Some time ago, scientists thought plate tectonics might be responsible for one of the most bizarre aspects of Mars' geology, something called the 'Martian dichotomy'. In essence, Mars' two hemispheres are completely distinct. The northern hemisphere sits an average of 5km (3 miles) lower than the southern hemisphere and the southern crust is significantly thicker. Because of this, Mars's northern regions are dubbed the lowlands and the southern part the highlands.

The difference in elevation between the two parts is stark, but the cause remains unclear. This dichotomy occurred extremely early on in Mars's history. It's possible that early Mars may have undergone a mega-impact in its northern hemisphere at the site of the Borealis basin, which is a large, smooth feature covering a staggering 40 per cent of Mars's northern hemisphere! This colossal collision may have blasted material from the northern hemisphere to wrap around the southern regions as a giant blanket of ejecta (the same huge impact may also have been responsible for forming Mars's two small moons).

On the other hand, multiple impacts may have occurred in quick succession to form the same basin and create the same blanketing effect – but that this would happen so many times in exactly the same region, and nowhere in Mars's southern half, would be unlikely. Tectonics is another possibility; continuously shifting, faulting, flexing plates could have created an uneven dichotomy in the Martian mantle. Again, while this is possible, we still have no proof of tectonics on Mars. Another idea is that some internal process caused the lowlands to sink and drop down, potentially an asymmetrical convection process that affected the mantle differently in different areas (such as underground plumes shooting upwards in one hemisphere and downwards in the other).

The two hemispheres aren't just different in terms of elevation. The planet looks like two different worlds haphazardly stuck together: the southern highlands are incredibly ancient and cratered, while the northern lowlands appear to have undergone a facial! The northern crust is younger, thinner and the surface is smooth, far smoother than the highlands. Some believe it to be a giant, sunken, hemisphere-wide sea floor, while others suggest that it has been extensively resurfaced by lava flows and volcanic eruptions over time (it may be a combination of the two).

This difference is evident in photographs of Mars. The cratered, pitted southern half wraps around the planet, sitting higher in relief than the smoother half. The separation is somewhat muddied by Mars's various albedo features. The planet is a mottled swirl of reds, browns and blacks, with lighter 'continent'-like regions standing out from darker 'seas' (and this was indeed how they were initially named, as lands, peninsulas, seas, bays, straits).

One of Mars's most obvious dark albedo features – actually a low-lying shield volcano – is named Syrtis Major. Its darkness is caused by both a relative lack of dust and the composition of the rock (volcanic basalt). The most notable string of darkness on Mars's surface (named Sinus Meridiani) begins at the eastward end of Valles Marineris and patchily makes its way over to Syrtis Major before snaking onwards. Other notable dark patches include Acidalia Planitia, perhaps the most obvious patch in the northern hemisphere, and parts of Noachis Terra (southern). These dark and light features change over time, partially driven by seasonal winds sweeping up surface dust and moving it around.

Last up is the Amazonian period, which stretches from roughly 3 billion years ago to the present day. While the Hesperian period was one of change, in which running water disappeared and Mars's atmosphere thinned dramatically, the Amazonian saw cold conditions really take hold and dominate

Martian geology. This period is characterised by its coldness, its aridity: in essence, it's the birth of Mars as we know it.

The period is named after Amazonis Planitia, a young lava plain with very little cratering. In fact, very few lava plains on Mars are quite as smooth as Amazonis Planitia, making it quite unique. It visually resembles some of the most barren and icy landscapes we see on Earth. Some have likened it to the terrain of Iceland due to its alternately smooth, streamlined, wrinkled, ridged, sloped and fractured appearance. While Amazonis Planitia is anomalously smooth, plains from this age are generally smooth and marked by only small numbers of craters from relatively recent impact events.

Mars's geological history is quite different to that of Earth. As mentioned before, Mars formed early on in the history of the Solar System. This means that many of the earliest impacts are actually preserved on some sort of crust. If the same impactor(s) had struck Earth, there would have been no crust to strike, as our planet was still very much within the molten part of its formation process. Additionally, Mars's entire history of water came to an abrupt end some 3 billion years ago. While we do have evidence of water and ice existing within the Amazonian period, it is a very different animal to the water of the Noachian and early-Hesperian periods: it is trapped in seasonally melting groundwater and briny flows, in the polar caps and potentially in underground ice sheets or aquifers.

Earth has many geologically dynamic processes occurring on its surface, such as tectonics and longer-term erosion and weathering (facilitated by our atmosphere), which have kept the surface of our planet changing over time. Mars, however, is largely unchanged. This is exciting to scientists. Although we know that present-day Mars is unlikely to be habitable, or could likely only be inhabited in anomalous or seasonal circumstances, it may once have been able to host life. Any leftover signs of life may be relatively easy to unearth, held frozen in time.

One vital thing missing from this history of Mars is magnetism. Today, the planet only shows weak signs of magnetism. Most of the planets in the Solar System have magnetic fields and thus magnetospheres (the region of space over which a planet's magnetic field – which extends into space like a giant bar magnet's – dominates). When a planet lacks a magnetosphere, it essentially lacks a shield against cosmic radiation and the intense solar wind, the stream of charged particles continually flooding out into space from the Sun.

These magnetic fields are generated in a planet's core, by the constantly heating-rising-cooling-sinking motions of molten metal (which acts as a giant cosmic dynamo). Earth has a solid inner core wrapped in a molten outer core; the inner core slowly grows by crystallising and solidifying, passing heat to the outer core as it does so to help sustain the ongoing convection currents there. Mars's core has a radius of 1,500 to 2,000km (930 to 1,240 miles); various observations have suggested that this core is at least partially molten, but we're unsure of its exact composition.

Whatever its core is like, Mars lacks a notable magnetosphere and is suffering for it. Its surface is constantly bombarded with radiation, and it's thought that the solar wind is at least partially responsible for stripping away the once-thick Martian atmosphere. A magnetosphere would have kept the incoming winds in check and protected at least part of Mars's atmosphere, but the planet no longer has a magnetic defence system.

The only two planets in the Solar System without a significant magnetosphere are Venus and Mars. Venus rotates incredibly slowly (it takes 243 Earth days for the planet to rotate just once on its axis!) and we thus don't believe the material in its core to be rotating very fast or to be convective overall, making the planet's lack of magnetic field somewhat understandable. However, given its characteristics, we'd expect Mars to have one. The planet's current field strength peaks at around 1,500nT (nanotesla); for reference, Earth's can reach around 65,000nT in strength. There is a key difference between the dominant magnetism seen on Earth and on Mars. Rather than boasting a magnetic field generated by a core dynamo, as our planet does, modern-day Mars has magnetism that is located in sporadic patches scattered across the planet, primarily in its southern hemisphere (these magnetised regions are known as 'crustal fields'). The fact that the rock forming Mars's surface is magnetised implies that Mars did have an intrinsic field at one point in its history, even if it is no longer present.

So, why did it switch off? Scientists have a few ideas. One possibility is that Mars was violently smashed into by a large impactor when it was just a few hundred million years old, heating its surface, disrupting its core dynamo and dismantling its magnetic field (also draining its atmosphere as a result). This is possible, as the Late Heavy Bombardment was in full force at that time in Mars's life, and we have other signs that Mars may have suffered a colossal impact (the Borealis basin, its moons, the crustal dichotomy). Scientists have matched age estimates for magnetised crater basins on Mars to the time when we believe Mars's magnetic field switched off for good, supporting this theory.

Such a collision may not have been destructive enough – but that might not matter. It's possible that Mars's dynamo simply decided to stop working on its own.

If it turns out that Mars's core lacks a large enough solid inner component to help heat an outer molten layer, the convection currents there may have stopped circulating. Even if that were not the case, Mars is simply smaller than Earth, so it may have run out of energy to self-sustain its interior convection anyway. If you take two spheres, one roughly half the size of the other, the smaller sphere has a far larger surface area relative to its volume. The larger the surface area, the larger the amount of heat that a body loses. For Mars, this meant the planet cooled far faster than Earth, potentially causing its core metals to stop their circulating motion.

Whether it died violently or not, we know that the magnetic field shut off long ago, leaving just traces of remnant magnetism to show for it.

CHAPTER SEVEN

Phobos and Deimos

The Solar System is chock-full of moons. It contains over 180 of varying shapes, sizes, compositions and colours. Some are peppered with erupting volcanoes and liquid pools of sulphur, others are slowly being ripped apart by their siblings' or parent planet's gravity, some are marked with angry red scars, and yet others are encased in thick layers of ice, some of which are suspected to rest atop vast, as-yet-unexplored subsurface oceans.

Despite the vast diversity of moons in the Solar System, Mars's little family manages to stand out. Mars is orbited by two potato-shaped moons: Phobos, which is 22km (13.5 miles) wide, and its little brother, Deimos, which is 12km (7.5 miles). Both were discovered in 1877 and named after characters from Greek mythology representing fear, terror, panic and dread. The names might sound unnecessarily extreme, but there's a good reason for them: as mentioned in previous chapters, Phobos and Deimos were the sons of Ares, the Greek god of war and counterpart to the Roman Mars.

For as long as we've been launching spacecraft towards Mars we've also been trying to learn more about the planet's two satellites, exploring everything from their shapes, sizes, densities and orbits, through to their composition, interior structure and surface properties, and postulating on how they both formed, have evolved and will ultimately die.

Numerous probes have managed to photograph either Phobos, Deimos or both, including *Mariners 7* and *9* (launched in the 1960s), *Vikings 1* and *2* (1970s), *Mars Global Surveyor* (1990s and 2000s), *Mars Reconnaissance Orbiter* (2000s), the *Spirit, Opportunity* (2000s) and *Curiosity* (2012 onwards) rovers, *Mars Express* (launched 2003) and others. The rovers have managed to view each moon passing in front of the Sun, and *Curiosity* observed Phobos passing in front of its little brother Deimos (a transit) in August 2013.

In 1988 the Soviet Union launched a dual mission dedicated to Phobos, named *Phobos* (or *Fobos*) *1* and *2*. Sadly, *Phobos 1* failed before achieving its aim and *Phobos 2* was only slightly more successful. Although we lost contact with the latter orbiter before it was able to send a lander down to the moon's surface, *Phobos 2* did manage to send back nearly 40 images of Phobos's surface. Russia tried again in 2011 with *Phobos-Grunt* (or Fobos-Grunt), a sample return mission destined for Mars's largest moon – but this mission failed to leave Earth's orbit, eventually falling back to our planet in January 2012 and landing somewhere off the coast of Chile.

Despite our limited data, we know that Phobos and Deimos are very different from our own moon – but we're unsure of exactly how and why that's the case. Like father, like sons!

Other than their current orbital characteristics, there's relatively little we know about Mars's satellites for certain. Although we have mapped the craters and features seen on their surfaces, probed their compositions, studied their spectra, modelled the ways in which they might have formed and evolved over time, characterised their orbits and snapped higher and higher resolution photographs of them, we still have many questions to answer. For example, how did Phobos and Deimos form? How are the two related to one another – if at all? Could there be water ice buried deep inside? What created their various surface features? What are they made of?

To start with, the moons' properties sit somewhere between those of a moon and an asteroid. Neither Phobos nor Deimos is massive enough to have pulled itself into a spherical shape, so they look more like lumpy, bumpy, knobbly rocks.

Deimos in particular has a very irregular shape. In fact, most of the moon's southern regions and south pole are completely missing! We think that the lower parts of Deimos were scooped out by a shattering impact with another body. Over time many of these jagged edges have smoothed out and some of the gaps filled in, likely due to debris flung up by the devastating crash itself falling back on to Deimos and re-blanketing it with material. With a rough diameter of 10km (6 miles), Deimos's southern concavity is possibly the largest recognised crater with respect to the host body's size (12km or 7.5 miles)!

This isn't the only sign of disruption on Deimos. The moon is reasonably cratered, with a couple of the largest ones – named Voltaire and Swift – measuring roughly 1 to 2km (0.6 to 1.25 miles) across. However, its overall surface is surprisingly smooth, as a large amount of the porous, fine and grainy 'sand' or 'soil' coating the moon's surface, known as 'regolith', has trickled down into any existing craters, filling them up and levelling out large areas of the previously rough surface. This regolith moves down the moon's slopes, stretching out to form elongated streams and streaks. Overall, Deimos resembles a small, oddly shaped pebble, with a polished surface that's sporadically pitted with craters.

Visually, Phobos couldn't be more different to its smaller brother. Phobos's entire surface is rough and heavily cratered; compared with Deimos's smoothness, Phobos resembles a cosmic pumice stone, a chipped and pitted lump of rock that's endured a real battering from meteoroids and other small bodies. Many of Phobos's craters are in different stages of degradation – some have very well-defined rims, while others have been worn down over time. The most prominent crater is Stickney, which stretches out for over 9km (5.5 miles) and completely dominates the small moon's shape.

As an aside, the naming of features on the planets and moons of our Solar System is overseen by the International Astronomical Union (IAU), which sets guidelines for each body. For Deimos, feature names are limited to those of 'deceased authors who wrote about Martian satellites', and for Phobos those of 'deceased scientists involved with the discovery, dynamics, or properties of the Martian satellites, and people and places from Jonathan Swift's *Gulliver's Travels*.' This is demonstrated in Deimos's only two named features – Swift and Voltaire (the chosen pen name of French writer François-Marie Arouet) – and in Phobos's most famous feature, Stickney crater (named after Angeline Stickney Hall, wife of the discoverer of both Phobos and Deimos, Asaph Hall).

Stickney is roughly 1km (0.6 miles) deep and filled with a mix of debris and regolith. The impact that formed Stickney stole a huge chunk of material from Phobos, and may have come close to shattering the moon entirely. Debris flung out by the initial impact formed a further scattering of smaller impact craters, peppering Phobos with dents, bumps and dimples. Like Deimos, Phobos is also covered by a waist-deep layer of dust, likely created by ongoing meteorite impacts that have flung material around and slowly chipped away at the moon's shape.

The surface features on both moons have undergone varying levels of space weathering – processes that not only wear down physical features, similar to geological weathering here on Earth, but that change a feature's optical properties in terms of darkening, reddening, levels of mineral absorption and more. This can be due to ongoing impacts from physical objects such as micrometeoroids, or due to exposure to radiation from the Sun and incoming cosmic rays. Signs of space weathering are evident in rock and soil samples returned from the Moon via the Apollo missions, and we see the same kinds of optical effects on both of Mars's moons.

The most noticeable feature on Phobos is also one of the most puzzling. Across its surface is a prominent and widespread network of scars, as if a set of claws frantically clawed and scratched at the moon and left deep gashes behind. We first spotted these marks via the Viking missions of the late 1970s.

These depressions and grooves are some 100 to 200m wide, tens of metres deep and stretch out for many kilometres in length (320 to 650ft wide, 100ft or so deep and several miles long). They are seemingly arranged in groups or families, with members of the same group parallel to one another. Most of the moon's surface is marked with these mysterious lines – they slice through a range of other features on the surface, including raised crater rims, with ease. Many run very close to one another, almost shoulder-to-shoulder, while some are narrow, some are wide and some are shallow. Some run uninterrupted for kilometres, while others are stubby and patchy. The longest groove is around 30km long (18.5 miles)!

Many of the grooves display 'pit chains', which are exactly as they sound: strings of small potholes and pits running along the length of the marks themselves. Pit chains have also been found on the surface of Mars, as spied by ESA's *Mars Express* spacecraft, and on the Moon and the Earth, most prominently in Hawaii and Mexico. Pit chains can be caused by stresses in the body's crust, volcanic activity and – the most exciting theory for possible microbial life on Mars – rock collapsing into hidden pools of groundwater. Many other mechanisms have been suggested for the formation of these pit chains on Phobos, from erupting jets of gas to rocky boulders bouncing downhill.

Deimos shows no signs of the odd scars present on Phobos. While we have seen similar features on other small bodies in the Solar System, namely asteroids and small icy moons, none are quite the same as what we see on Phobos. While we do have a few ideas about how Phobos's grooves formed, they're still somewhat mysterious.

Scientists think it likely that multiple mechanisms are at play with Phobos's grooves, rather than there being a single cause. Initially, we believed the marks to emanate from the region surrounding the Stickney crater, giving rise to the idea that they formed in either the initial impact or the subsequent rain of debris caused by the collision. However, recent studies have scrapped the idea that the grooves are intrinsically related to Stickney; the visual relation was simply coincidental. Rather than originating in the Stickney impact zone, these marks are now thought to extend from a point near to the crater instead.

It's possible that other impacts on Phobos threw up batches of pebbly debris that later rained down on the moon, or that debris from impacts on Mars behaved similarly. However, it's difficult to reconcile the precise parallel arrangement of the grooves with such processes; while these mechanisms could likely form some of the grooves on Phobos – some of the wider and more distinct grooves, for example, or some of the orientations and morphologies we see – they probably couldn't be responsible for all of them. It's also possible that any existing debris whirling about in orbit around Mars could have impacted Phobos, but this idea suffers from the same problems. Additionally, there's a single patch of Phobos's surface that is surprisingly clear of grooves, which wouldn't make much sense if debris was striking Phobos indiscriminately like ammunition from a machine gun, carving out depressions as it skimmed across the moon's surface. If that were the case, it's probable that the entire moon would be evenly covered.

Another option is that fragments and hunks of rock thrown off by impacts on Phobos in the past have rolled around, steamrollering across the small moon and leaving prominent tracks in their wake. While the intuitive nature of this option is appealing, the problems are evident: why would rolling objects create consistently parallel patterns that, rather than being deflected by a crater wall or rocky outcrop, appear to cut right through existing features with no apparent problem? We'd expect moving ejected material to meander and be less consistent in its behaviour.

Over the past few years, another theory has gained attention. This proposes that Phobos's grooves may not have been created by objects hitting or digging into the surface. They may actually have been created by the disruptive influence of Mars!

This is certainly a possibility if you examine Phobos's unique situation. Of all the moons in the Solar System, Phobos is by far the clingiest. It orbits just under 6,000km, or 3,728 miles, from Mars's surface – closer than any other moon in the Solar System. This is so close that if one were to stand in Mars's polar regions, Phobos might actually be obscured from view by the curvature of the planet itself! Deimos orbits just over 20,070km, or 12,470 miles, from Mars's surface – somewhat further away but still by no means distant. For comparison, the Moon lies hundreds of thousands of kilometres away from Earth.

This extreme orbit has severe consequences for Phobos and defines much of its behaviour. For one, it causes a curious phenomenon that, again, is unique to Phobos. Phobos orbits Mars far faster than the planet itself rotates. The moon zips around Mars in just over 7.5 hours, while Mars itself rotates once on its axis in just over 24.5 hours. This means that Phobos orbits Mars three times in one Martian day! In just a single day, the moon rises and sets in the sky multiple times (at least twice, sometimes three times). Comparing this with a more familiar situation highlights just how unusual it is: while Earth rotates roughly once on its axis every 24 hours, the Moon takes a full 27 days to complete one orbit. As Phobos's discoverer Asaph Hall commented wryly, even 'the most zealous astronomer might tire of observing a moon three times in a day'.

This has a weirder consequence than you might think. Our Moon rises in the east and sets in the west. Phobos, however, appears to do the exact opposite.

Because Phobos is moving so speedily, from the Martian surface it appears to rise in the west and set in the east. This can be a tricky concept to grasp, but it helps to consider it from Mars's perspective. As a (very crude) thought experiment, imagine stargazing on the surface of Mars, but with your body positioned at the centre of the compass points (west to your right, east to your left). As Mars rotates in the same direction as Earth (anticlockwise as viewed from the planet's North Pole), the sky appears to 'rise' in the east and 'set' in the west. As Phobos is continually overtaking Mars in its orbit, it zips around the planet and reappears in the sky in the west, moving quickly across the sky and disappearing in the east. If you do the same thought experiment for a moon that rises and sets just once per day – in other words, a moon that orbits slower than the rotation of its planet, like ours – it yields the opposite result, as the planet is the one doing the 'overtaking'.

A single day on Phobos, the amount of time taken for the moon to spin once on its axis, lasts for exactly the same amount of time the moon takes to orbit Mars once, something known as 'synchronous rotation'. As Phobos orbits Mars, it spins at the precise rate needed to ensure the same side is always facing Mars. This is due to a phenomenon known as tidal locking and means that any residents of Mars would only ever see one side of Phobos. The same is true for Phobos's little brother Deimos, which zips around Mars in 30.3 hours. While this by no means as extreme as Phobos, the length of Deimos's orbital period is close enough to that of a Martian day (under six hours longer) that it would rise and stay in the sky for several days before setting.

We also experience the phenomenon of tidal locking with our own Moon. Although we can get glimpses of just over half of the Moon's surface over extended periods of time (a process known as 'libration', which Martians would also experience when observing their moons), we only ever see the 'light' side. Tidal locking is caused by the gravitational effects of the primary body (in this case, Mars) distorting and subtly reshaping a moon (Phobos or Deimos) over time such that the moon's rotation rate also changes. The majority of the main moons in the Solar System are tidally locked. Tidal effects become more significant when moons are both small in comparison to their parent planet and orbiting relatively closely (both true for the Martian system).

Phobos is slowly approaching Mars by a couple of metres (6.5ft) per century. This inward spiral could literally be the death of Phobos – in time it'll drop below something known as Mars's Roche limit, the altitude below which objects are torn apart by the planet's gravity and tidal forces. For the Mars–Phobos pairing, this is not that far below Phobos's current orbit. Before 50 million years are up, Phobos will be in real danger.

Deimos escapes much of this hassle due to its greater distance. While Phobos is spiralling inwards, Deimos is slowly moving away from Mars, although far slower than the Moon is receding from the Earth. Deimos and our Moon are slowly moving outwards because they complete one orbit slower than it takes their host planet to spin once on its axis. Because of this, and the resulting ways in which energy is exchanged between a planet and its moon, they are effectively spiralling away from their planets. Conversely, Phobos orbits quicker than Mars rotates and so is slowly winding closer to Mars.

It's possible that Phobos's extreme orbit is also responsible for the moon's grooves. Recent studies have suggested that these surface scars might be the cosmic equivalent of stretch marks, created by Mars's gravity tugging and stretching on Phobos, subtly changing its shape and deforming it over time. This idea is supported by the fact that some of Phobos's marks appear to be younger than others, overprinting more recent impact craters and their ejected material. The gravitational tug-of-war between Mars and Phobos might have given rise to stress fractures.

'We think the grooves are signs that this body is starting to break apart tidally, and that these are the first evidence of the tidal deformations of Phobos,' said planetary scientist Terry Hurford of NASA's Goddard Space Flight Center in Maryland, US, in 2015. 'Eventually, Phobos will be ripped apart before it reaches Mars's surface.'

Variations of this theory include mention of Phobos's internal structure. If the aforementioned tidal stretching creates fractures and stretch marks across Phobos, the moon's regolith might drain down into its interior, as sand trickles through an hourglass. Alternatively, the regolith could be ejected and blasted away by various gases escaping and streaming out through the fractures. This outgassing might be driven by hidden stores of water ice locked deep down inside Phobos.

Whether or not this model is correct, it highlights important, and currently unresolved, questions about the interior structure of Mars's moons: what are the two moons made of and what are they like inside?

Several decades ago we thought that both moons might be archetypes for small Solar System bodies, but that's turned out not to be the case at all. As mentioned previously, both are oddly shaped potatoes. Phobos is slightly elongated along one axis, most probably due to the influence of Mars stretching it out like toffee.

If you were to stand on the surface of either moon you'd experience much reduced gravity: Phobos's and Deimos's surface gravities are just under 2,000 and 5,000 times weaker than Earth's (0.0055 and 0.002 m/s^2, respectively). In order to break free from Deimos's gravity, for example, you'd only need to be travelling at around 20kmph (12.5mph) – a pretty average throw of a baseball would send it hurtling free, effectively creating a new Martian mini-moon!

We currently think that both Phobos and Deimos are little more than tenuous piles of cosmic rubble, despite their apparently cohesive surfaces. We suspect this because they appear to have very low densities (1,860kg/m^3 for Phobos and 1,490kg/m^3 for Deimos*), implying that their insides are very fractured and weakly held together. It's possible that the two moons could contain lumps of water ice frozen deep down beneath their surfaces. Essentially, their densities are far too low for the moons to be composed of solid rock. Both moons are also substantially less dense than Mars itself, meaning that they must either be very porous inside, with large cavities sitting between regions of rock, or formed from material that is inherently far less dense (the latter conclusion would have strong implications for how the moons formed in the first place) – or a mix of both.

---

* For comparison, Earth has a density of 5,514kg/m^3, and the Moon of 3,344kg/m^3.

Studies suggest that both moons have much in common with outer-belt asteroid material, namely either C- or D-type asteroids. C-type asteroids are otherwise known as carbonaceous chondrites – this group of stony, non-metallic meteorites includes some of the most primitive meteorites known to exist. These crude objects are characterised by the high proportion of little round grains (chondrules) they contain, thought to be the very earliest building blocks of the Solar System. As their name suggests, carbonaceous chondrites contain carbon compounds and amino acids, and often show signs of water. D-type asteroids are not dissimilar; thought to have come from the outer Solar System, these asteroids are very dull, reddish, and unreflective, and are thought to contain silicates (rich in organics) and possible water ice.

Both types of asteroid are generally quite unreflective and reddish in colour, something that strengthens the case for Phobos and Deimos having a similar composition. Both of Mars's moons are incredibly dark – they are some of the darkest known bodies in the Solar System. They have low albedos, meaning their surfaces don't reflect much of the light that falls on it. Albedo ranges from 0 to 1 from least to most reflective; a body with an albedo of 1 is a perfectly reflective body. Fresh snow, for example, can have an albedo of up to 0.9 (meaning it reflects 90 per cent of the incident light), while something like tar/asphalt is somewhere around 0.04 (reflecting around 4 per cent).

Really dark astronomical surfaces are mostly dark due to a combination of initial composition, which would usually be somewhat primitive, and ongoing space weathering, which can darken a surface over time. Phobos has an albedo of around 0.071 and Deimos around 0.068 – very similar and very dark. The darkest known asteroids in the entire Solar System have albedos of around 0.05.

As well as telling us a lot about what the moons are currently like, and how they'll continue to evolve, knowing more about the compositions of the two moons is key for understanding how they formed in the first place. There are a few main proposals for the formation of Phobos and Deimos: the moons could be native to the Martian system and have formed in situ around the planet, they might have formed elsewhere in the Solar System and only been adopted into Mars's cosmic family later on, they might have formed via a giant collision, or by some combination of these scenarios – or via another process completely.

The birth of Phobos and Deimos

Firstly, and perhaps most famously, is the idea that Phobos and Deimos aren't really related to Mars at all: they're actually adopted asteroids that were captured by the planet at some point in its history. Mars may have managed to snag a couple of single nearby objects, pilfered one of the objects in a duo as the pair flew past (this would have had to happen twice, once for Phobos and once for Deimos), or grabbed a single object that later split into two post-capture. The objects captured by Mars could have been from the inner or outer Solar System or from the main asteroid belt; each of these scenarios would result in a different composition for Phobos and Deimos, with a different internal structure and mix of water, organics and minerals.

Initially, this looks appealing; the moons' spectra do suggest that both bodies may have similar compositions to asteroids, and Mars does sit close to the asteroid belt. In order to be asteroid-like, the moons must have swept up material that lies far beyond Mars's orbit, implying that they didn't form where they currently live. However, there are a few issues with this, including the fact that capturing an intact object is actually a very, very unlikely event – and for Mars, it would have to have happened not once, but twice.

The behaviour of the moons themselves throws up additional hurdles for this theory. Both have properties that are difficult to reconcile with them being captured asteroids. Both Phobos and Deimos have orbits that are very circular (in fact, almost perfectly circular), prograde (meaning they orbit in the same direction as Mars rotates, anticlockwise, like the vast majority of bodies in the Solar System) and very closely aligned with Mars's equatorial plane (known as 'low inclination' orbits).

These orbits are very difficult to explain if we assume the moons to be snagged passers-by. Such bodies would likely be disruptive and travelling on any number of differently aligned orbits. An incoming asteroid could approach from a range of directions and inclinations. In order for the current orbits of Phobos and Deimos to be so well behaved, they must have been altered, reshaped, realigned and circularised over time. This would have required Mars to have a very thick primitive atmosphere, not only to considerably alter the moons' orbits, but so that the planet was capable of removing the excess energy from the passing bodies after it hooked them to slow them down.

Crucially, this thick atmosphere would have had to dissipate rapidly at just the right time in order to avoid either ripping the new moons apart or pulling them down to crash into the surface of Mars. While this is all possible, it requires some precise timings and is therefore unlikely, particularly for Deimos. Phobos is so close to Mars that atmospheric drag could potentially have had this kind of effect, but Deimos exists much further away from Mars. To apply to both moons, Mars's super-thick primitive atmosphere would thus need to have been far-reaching, too.

A good example of the capture mechanism is Triton, Neptune's largest moon, which is a captured dwarf planet from the distant outer Solar System (a body similar to Pluto). However, Triton does not behave as well as Phobos and Deimos do; it orbits Neptune in the 'wrong' direction (retrograde, opposite to Neptune's rotation) and at a very high inclination. It likely disrupted the entire Neptunian system when it joined the family, throwing some of Neptune's other moons into disarray and potentially flinging earlier moons out of the system completely.

It's doubtful that Mars could have captured two passing bodies and reshaped their orbits so significantly – scientists dub it anywhere from 'almost impossible' through to 'extremely improbable' and 'unlikely' – but it remains a possibility.

A second idea proposes that the moons simply formed alongside Mars. As Mars grew larger and gathered material from its surroundings, so did Phobos and Deimos. The two forming moons whirled around their parent planet, eating up material in their path just as Mars did. This would accurately recreate the moons' neat orbits and good behaviour, and could create the suspected rubble-like interior structures.

However, the major problem with this theory is the composition of both Phobos and Deimos: they're both very different to Mars, and if they'd co-accreted from the same material, they should instead be similar. Additionally, it's not known whether there would have been enough mass available around Mars to form both moons (but it's possible that a nearby collision could have dumped a load of mass into the planet's orbit, flooding the region with new building blocks).

Lastly, it's possible that the Martian system has a great deal in common with our own. While our knowledge of how our moon formed is by no means certain, the leading theory is known as the 'giant impact hypothesis'. This theory proposes that a Mars-sized lump smashed into Earth when it was still forming. This impactor, dubbed 'Theia' (after the mother of Selene, the Greek goddess of the Moon), dealt Earth a destructive blow, sinking deep into the young planet and causing devastation. Material from both Theia and Earth was flung out into space, settling into orbit around Earth. Over time this debris came together and combined to form the Moon. Earth also gathered up some of this material, wrapping it around itself as a new veneer. This theory creates a neat circular orbit for the Moon, which is the Achilles' heel of capture theories.

Something similar may have happened with the moons of Mars. Early in Mars's life, a giant Moon-sized body could have violently collided with the planet, shattering into pieces as a result. Material from both the impactor and Mars's early crust would have been thrown out into space to form an orbiting disc, stretching out to beyond the present-day orbit of Deimos. The moons could have then formed within such a disc, hungrily gathering up rocky building blocks and growing larger over time.

There are a few constraints on such an event, including the size of the collision (large), the overall mass contained within the disc (greater than that of Phobos and Deimos combined) and the size of the disc (stretching out to the current orbit of Deimos), but many of the signs look promising. A wide range of impactor orbits could have formed such an accretion disc, but the impactor likely had a mass of around a few per cent of Mars's mass, forming a disc stretching out towards Deimos (a larger disc would have formed moons beyond Deimos, which we don't see). This disc would have contained far more mass than is currently constrained within Phobos and Deimos, and thus would likely have formed a number of large inner satellites that have already spiralled inwards to collide with Mars.

Additionally, there are several different impact basins on Mars that could be related to such a collision, such as the Daedalia, Chryse and Borealis basins, and parts of the highlands. Such a collision could explain why Mars's two hemispheres are so disparate, with the southern half sitting much higher in elevation than the northern: much of the material from the northern hemisphere may have been lost to space and used to form Mars's twin sons.

One of the current issues with this hypothesis for the Earth–Moon system is that of the Moon's composition. Our satellite is very dry and lacks many of the volatiles (substances that have lower boiling points, such as water and carbon dioxide) and minerals we'd expect to find in a body formed of similar stuff to the Earth. The Earth is remarkably watery to have such a bone-dry satellite! If the impact blasted some of Earth's bulk out into space, why does the Moon not have more in common with its parent planet?

Some suggest that the collision would have thrown up only a small proportion of Earth material, and would mainly have shattered the impactor itself into pieces, leaving a disc of predominantly impactor-sourced material to coalesce into the Moon. As for the Earth-sourced material, all of the volatiles could essentially have been baked off into space by the extreme heat generated by the original collision. Such a disc could explain the apparently water- and volatile-poor compositions of both of Mars's moons, too. The disc would have been heated in its location around Mars, baking the water, mineral and volatile content out of the forming moons.

This hypothesis has gained significant support in recent years and many believe it to be the most likely explanation for the formation of Mars's moons. Recently (July 2016), researchers combined numerous models for the formation of the moons and concluded that Phobos and Deimos most likely formed from an impact, as described above.

According to their models, a forming planet smashed into Mars within the first billion years of its life and threw huge amounts of debris out into orbit around the Red Planet, shattering into pieces as it did so. This ring formed a number of initial – and larger – moons orbiting close to Mars, which dynamically interacted with the remaining dusty, pebbly little building blocks surrounding them and influenced the material there to start clumping together to form the potato-like duo we see accompanying Mars today. These initial moons were eventually dragged inwards to be torn apart and consumed by Mars, and Phobos and Deimos slowly moved and repositioned themselves in space, ending up in their current locations billions of years later. This would explain why the two moons have a different composition to their 'parent' planet – as seen with Earth and the Moon – and manages to successfully reproduce the Mars system as we see it today (which is no mean feat).

If things weren't uncertain enough, it's also possible that one of these theories might apply for Phobos and another for Deimos. As with so many things concerning Mars, we are not really certain about anything; to whittle down the list of 'maybes', we simply need to know more.

Such knowledge will come once we successfully send spacecraft to directly explore (and perhaps land on) the moons themselves. NASA has actually raised the possibility of landing on the larger of the two moons, Phobos, before landing on Mars itself; the moon is far closer to Mars than Earth and so could communicate swiftly with any technology on the surface, landing in and launching from lower gravity is likely to be less demanding fuel-wise (even if not easier, with the higher bounce risk) and it would be an ideal base from which to observe its parent planet given its speedy path around Mars. Logistics aside, it would be a colossal step forward in our understanding of the entire Martian system.

CHAPTER EIGHT

The Draw of Cydonia

When you hear the name Nikola Tesla, what springs to mind?

Those in the know may think of the physicist's (1856–1943) association with famous inventor Thomas Edison, Tesla's various inventions and discoveries – the most famous of which involve electricity and magnetism via his work on alternating current and the Tesla induction coil – or Wardenclyffe Tower (the Tesla Tower), a wireless radio station Tesla attempted to build and use for intercontinental, and maybe even interplanetary, communication in the early 1900s before he ran out of money.

Some may think of Tesla Motors, the electric car company now headed by the coincidentally Mars-obsessed Elon Musk. Tesla Motors's founders supposedly spent ages thinking up the ideal name for their forward-thinking business concept, before settling on Tesla as an appropriate namesake.

Interestingly, many may think again of David Bowie, who played Tesla in the 2006 film *The Prestige.* The film's director, Christopher Nolan, described Tesla as 'extraordinarily charismatic', an 'other-wordly, ahead-of-his-time figure', and instantly knew he wanted Bowie to play him due to the latter's 'slightly different sort of star quality'. Bowie originally turned down the part, causing Nolan to beg him to reconsider. 'It occurred to me [Tesla] was the original *Man Who Fell to Earth*,' Nolan wrote in an *Entertainment Weekly* article just after Bowie's death in January 2016. 'Bowie seemed to be the only actor capable of playing the part. He had that requisite iconic status, and he was a figure as mysterious as Tesla needed to be.'

Others may think of Tesla's famously eccentric nature. Along with Einstein and Feynman, Tesla has, perhaps unwillingly, become a poster-boy for the 'mad scientist' trope, that of a troubled and bizarre form of genius. Admittedly, Tesla did himself no favours here; he was well-known for being a 'tall, gaunt electrical wizard who slept only two hours a night', and a recluse who eschewed romantic relationships with people in favour of loving friendships with pigeons. It may be an urban myth, but he reportedly once proclaimed to love a pigeon 'as a man loves a woman', waxing lyrical on the beauty of his pigeon love's bright eyes and white plumage.

He was also a strong proponent of an inhabited cosmos – Mars in particular – and believed that extraterrestrial life in general was an absolute certainty. 'Of all the evidences of narrow mindedness and folly,' he wrote, 'I know of no greater one than the stupid belief that this little planet is singled out to be the seat of life, and that all other heavenly bodies are fiery masses or lumps of ice.'

Like many prominent scientists of the time, Tesla believed that highly intelligent and advanced Martians, far more mechanically advanced than us, were desperately attempting to communicate with Earth. In 1899, when Mars was at opposition and sitting relatively close to our planet, Tesla claimed to have detected something amazing: radio communication from Mars! From his 'Experimental Station' in Colorado Springs, US, he observed 'feeble planetary electrical disturbances which, according to [his] investigations, could not have originated from the Sun, the Moon, or Venus'. The odd static-y signals appeared to abruptly cease as soon as Mars set in the night-time sky – coincidence, or proof that they were streaming out from the surface of the Red Planet?

The unexpected and peculiar signals terrified Tesla to begin with. 'There was in them something mysterious, not to say supernatural,' he wrote a couple of years after the detection. He ruled out various other possible causes – 'the Sun, Aurora Borealis, earth currents' – and eventually settled on the idea of 'intelligent control', concluding that the signals couldn't have been accidental. 'The feeling is constantly growing on me that I had been the first to hear the greeting of one planet to another,' he said.

Over the following months and years, Tesla became increasingly confident that the 'inexplicable, faint and uncertain electrical actions' he initially recorded were in fact deliberate transmissions from some Tesla equivalent residing on the Red Planet. 'Further study has satisfied me that they must have emanated from Mars,' he wrote in a letter to the editor of *The New York Times* in 1909. 'All doubt in this regard will soon be dispelled.'

Tesla deciphered the radio signals as being numerical, a code he thought to be universal for all alien civilisations across the cosmos. He described it as being a consecutive code of numbers, signalling one-two-three or one-two-three-four repeatedly. Beep, beep-beep, beep-beep-beep, beep-beep-beep-beep, over and over again, in quick succession.

This may sound more science fiction than science, but many scientists agreed with Tesla, supporting both his supposedly Martian radio signals and the idea of intelligent Martians overall. Many also remained sceptical, still believing the Martian atmosphere to be too weak to support life. Tesla himself was aware of the somewhat controversial nature of his discovery and knew that some would 'scoff at [his] suggestion or treat it as a practical joke', but remained convinced. He spent many years of his life seeking a way to communicate with other planets and striving to build a 'Teslascope' for this very purpose. Despite his scientific success in other areas – electricity, magnetism, X-ray imaging, radio, telecommunications, the list is enviable and almost endless – Tesla didn't manage this dream of interplanetary conversation, but he can hardly be blamed for failing to communicate with a civilisation that didn't exist in the first place.

Numerous other scientists also picked up on unexplained radio signals coming from outer space in the decades following 1899, and they were by no means mad. Many cosmic objects – alive and dying stars, galaxies, black holes, star-birthing clouds – emit radio waves that we routinely detect, but scientists weren't aware of this phenomenon until many years after Tesla's death. It's now thought that Tesla and his contemporaries were probably detecting radiation emanating from Earth's upper atmosphere (the ionosphere) or the Jovian system, or potentially even mistakenly tuning in to some of the radio experiments that were taking place in Europe at the time.

In 1924, a quarter of a century after Tesla's discovery, Mars entered another opposition: the twelfth, and closest, since Tesla's finding. This time, the US government decided that it'd be wise to be a little more prepared for the Red Planet's approach. With the cooperation of the US army, navy and government, an American astronomer – and friend of aforementioned *canali* advocate Percival Lowell – David Peck Todd sought the help of astronomers across the country by telegram, asking any willing participants 'who believe it possible that Mars may attempt communication by radio waves with the planet while they are near together' to report 'any electrical phenomenon of unusual character' between 21 and 24 August. This exercise, labelled 'National Radio Silence Day', returned no signs of Martian life, despite observers covering a wide range of wavelengths and numerous stations periodically going radio silent so as not to drown out or confuse any potential signals sent by conversation-hungry Martians.

However, there may be a legitimate reason for the silence from Mars. Martians may once have existed on the planet, only to be wiped out by some kind of catastrophe – or so said scientist John Brandenberg, who developed the so-called 'Cydonian hypothesis' in the 1990s in collaboration with colleagues Vincent DiPietro and Gregory Molenaar. Brandenberg's hypothesis claimed that Mars had once been inhabited by an indigenous race of highly developed humanoid aliens. These aliens, named Cydonians after the region of Mars in which they lived, built all kinds of intricate structures and carvings … including a giant, 2.5km-long (1.6 miles) humanoid face, painstakingly sculpted out of rock and gazing silently out into the cosmos.

The Face on Mars

In June of 1976, the *Viking 1* spacecraft completed its 10-month journey to Mars. It slotted neatly into orbit around the planet and began its two-step mission: the probe's orbiter was to photograph the planet from space and act as a communications link with Earth, while the lander would head down to Mars's surface and study the planet from below.

With the success of Martian rovers in recent years this may sound like a pretty straightforward mission plan, but *Viking 1* marked the first ever US attempt at landing on Martian soil. The Viking mission came to life following the Mars mission plans developed by NASA throughout the 1960s. NASA had initial plans to send a number of unmanned probes to the Red Planet in the mid-to-late 1970s under the name of Voyager*, but ended up scrapping them due to budget constraints and concerns over the launch plans. However, these plans weren't completely set aside, and resurfaced in a more affordable, scaled-back capacity in the form of *Viking 1* and *Viking 2*, launched on 20 August and 9 September 1975 respectively.

The *Viking 1* lander separated from its orbiter and arrived safely on Mars on 20 July 1976. Just minutes after landing, the industrious little robot began sending back photographs, beginning with a snap that was not only the lander's first, but the first ever sent back from the surface of Mars. *Viking 2* quickly followed in its sibling's footsteps, arriving in orbit around Mars in August 1976 and sending down a lander on 3 September. The four robots worked together to dramatically increase our knowledge about Mars and formed the basis of our understanding of the planet for the next few decades.

The probes found many surprising things on Mars, including water-hewn structures and features that indicated a far more watery past for the planet than we had anticipated. However, one region the Vikings explored attracted far more interest than the rest.

Mars is covered in a blend of albedo features that range from dark to light in colour. Sitting on the border between

* The name 'Voyager' was later used as a moniker for a couple of probes launched in 1977 to explore the outer Solar System.

one of the dark (Acidalia Planitia) and light (Arabia Terra) patches in Mars's northern hemisphere is a region known as Cydonia. Cydonia was photographed by both the *Viking 1* and *2* orbiters, and numerous images were sent back to Earth*.

One of Viking's photographs appeared to show something truly shocking – a giant humanoid face rising out of the Martian soil! Nearly 2km (1.2 miles) across, it had dark depressions for eye sockets, a small upturned nose, a dark line where its mouth should be and an oval face shape. Some even saw a headdress or helmet. It looked for all the world like a giant carving of a skywards-staring face.

NASA decided to mention this in the caption when they released the image, figuring that it would be a nice way to gain the public's attention. 'The picture shows eroded mesa-like landforms,' said the press release on 31 July 1976. 'The huge rock formation in the centre, which resembles a human head, is formed by shadows giving the illusion of eyes, nose and mouth.'

While the very wording of the press release dismissed any notions of a deliberately carved face – 'resembles', 'illusion' – many immediately became excited about the Face. It was lauded as proof of life on Mars, a remnant from an ancient civilisation, a crumbling statue that had been lovingly shaped from a rock by intelligent aliens. Some even saw some semblance of teeth in the mouth and a pupil in one of the eyes, features that would surely be proof of artificiality. The Face soon appeared everywhere!

One of the Face's key proponents was – and to a lesser extent still is – a man named Richard Hoagland, who strongly advocated that the mound didn't just *look* like a carving of a human face … it actually *was* an artificial structure created by Martians, a sculpture hewn out of the rusty soil, a depiction of a large face gazing at the sky. Brandenberg, the proposer of the aforementioned Cydonian hypothesis, agreed with Hoagland here; he believed the face to be 'a portrait of a

* These images achieved a best resolution of 43m (140ft) per pixel.

Cydonian' that happened to look pretty familiar to us because the inhabitants of Mars also happened to be humanoid.

Public interest in the Face was so strong that NASA decided to target the Cydonia region with their subsequent missions to the planet, stating that they felt it was 'important to taxpayers' to do so. NASA's *Mars Global Surveyor* (*MGS*) photographed the Face in April 1998, some 22 years after the initial Viking image that sparked the furore. The spacecraft's high-resolution photographs were ten times sharper than Viking's best image had been, and showed what scientists had expected all along: the Face wasn't actually a face. It simply didn't resemble one at all. It was an uneven mound with pits, wells, bumps, rocky ridges and inclines, that had been lit in such a way as to form a pretty good optical illusion of a face.

However, many remained unconvinced. When the newer *MGS* photograph was obtained the weather had been different, the angle and lighting wasn't precisely the same, and there had been haze that could have obscured the clear facial features from view. So in 2001, *MGS* targeted the Face once more and added yet another nail to its coffin.

The probe sent back an even better view of the mound, with a resolution of 1.56m (5.12ft) per pixel. The rough and cracked surface of the feature was even more evident, and it became clear that it was no different to the thousands of other hills in the area. These hills are known as mesas, and are ubiquitous across both Mars's surface and in arid places on Earth. Mesas are flat-topped mounds with relatively steep sides, created via weathering and rock erosion. Their very name, *mesa*, is Spanish and Portuguese for 'table', reflecting their characteristic table-top shape. A famous mesa on Earth can be found in Cape Town, South Africa, in the form of the aptly named Table Mountain.

MGS also carried an instrument known as the Mars Orbiter Laser Altimeter (MOLA), which pinged the Martian surface with pulses of light to figure out the height and relief of the Face. This technique is similar to echolocation, which is how bats navigate: by sending out parcels of sound, in their case, and listening to the changes in pitch and tone when they bounce off nearby objects and return to the ear. According to MOLA, the Face rose to an altitude of 240m (790ft) above the surrounding plains.

'We took hundreds of altitude measurements of the mesa-like features around Cydonia, including the Face,' said Jim Garvin, former chief scientist for NASA's Mars Exploration Program and co-investigator for *MGS*, in 2001. 'The height of the Face, its volume and aspect ratio – all of its dimensions, in fact – are similar to the other mesas. It's not exotic in any way.'

However, some – Hoagland in particular* – remained unsatisfied with the newer images, blaming everything from the lighting to the angle to the image processing. In his 2001 book *The Monuments of Mars: A City on the Edge of Forever*, Hoagland claimed the different lighting to have produced a '"ghoulish" result' akin to that created by a child illuminating their face from below with a torch on Halloween. 'Even a mother would have difficulty recognising her kid's grossly distorted resemblance under those conditions,' he wrote.

Rather than delving into conspiracy theory, others simply stubbornly and bizarrely believed that the higher-resolution images quite obviously still showed a face. They argued that the new data only served to confirm the 'architectural symmetry' of the mound, the subtle details of its facial structure (teeth, eyes, headgear), the proportions of the Face's 'features' (eyes, mouth, ridged nose) and three-dimensional structure … everything still aligned to show a face, despite the different angles and conditions used.

Some die-hard Face fans took the idea even further, suggesting that not only was the mesa a face, but it could

---

* Tellingly, Hoagland's publisher, Richard Grossinger at North Atlantic Books, labelled him 'a David Bowie-like creature, increasingly more prophetic and New Age, even loony … Hoagland is a unique mixture of amateur scientist, genius inventor, scam artist, and performer, blending true, legitimate speculative science with his own extrapolations, tall tales, and inflations. He is a brilliant and glorious myth-maker and [an] evidence-based scientist at the same [time]'.

Right: *Curiosity's* selfie (taken on 19 January 2016).

Below: Three generations of rover at NASA's Jet Propulsion Laboratory in California – a flight spare for *Sojourner* (front), and working test siblings for *Spirit* and *Opportunity* (left) and *Curiosity* (right).

Left: This 1976 *Viking 1* image shows Mars's Cydonia region, with the 'Face' visible just above centre. This batch of images had a best resolution of 43m (141ft) per pixel. Speckling is due to missing data.

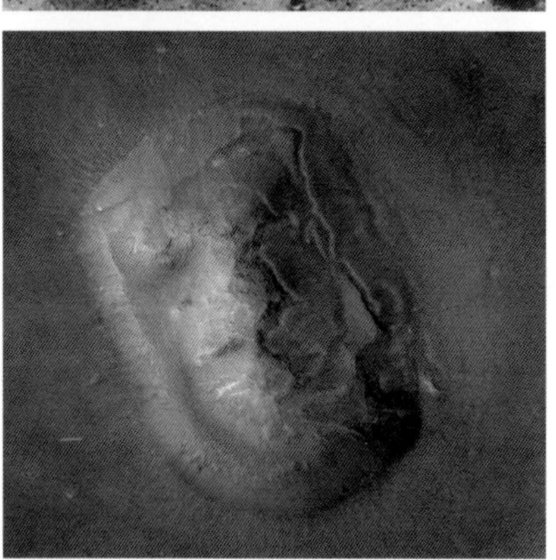

Left: Revisiting the 'Face'. This 2001 *MGS* image has a resolution of around 2m (6.6ft) per pixel, and revealed the 'Face' to be an unremarkable natural landform.

Below: A view of Cydonia (*Mars Express*). The 'Face' is visible in the centre, and the 'City' and 'D&M Pyramid' can be seen to the upper left. The image resolution is approximately 14m (50ft) per pixel.

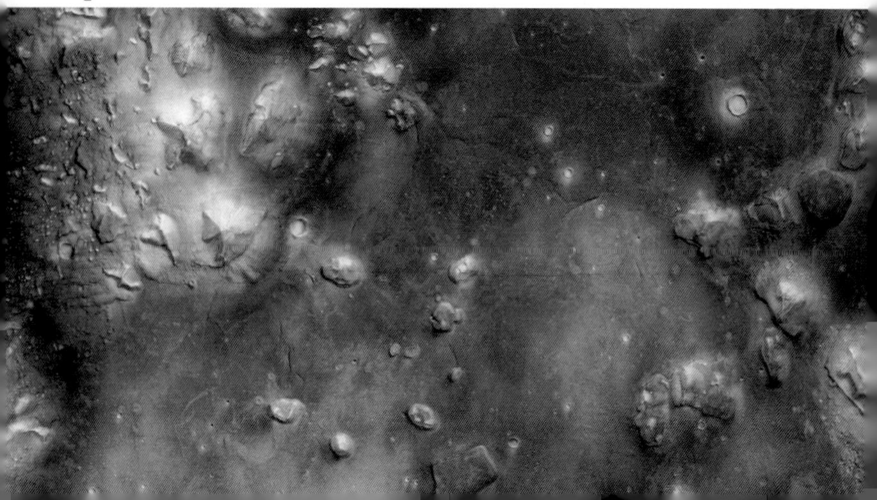

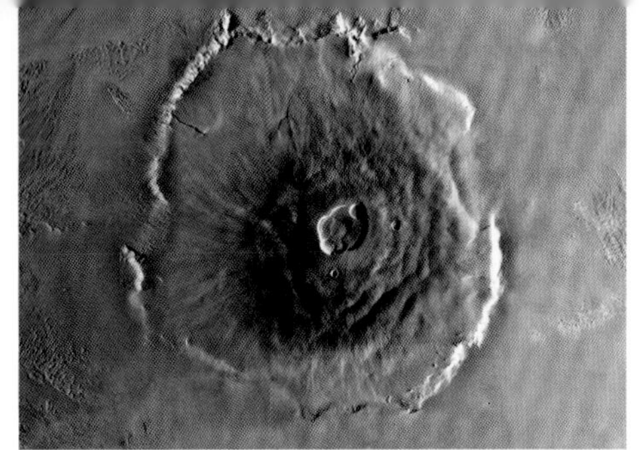

Left: This colour mosaic shows Olympus Mons, the largest volcano on Mars (*Viking 1*, 1978).

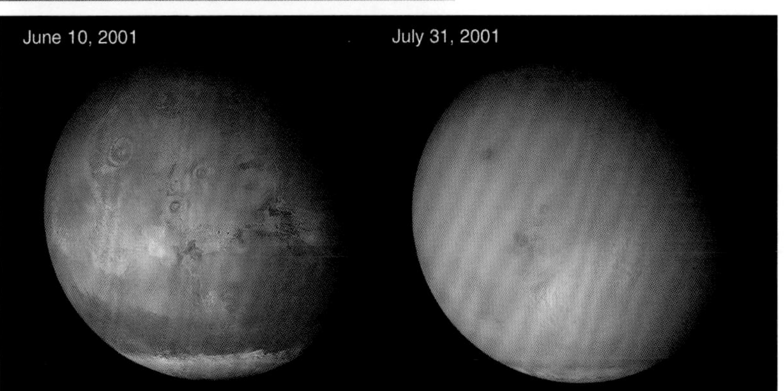

Above: These two *MGS* images show Mars before (left) and during (right) a dust storm in 2001.

Left: *Curiosity* used a new technique when it set wheel on Mars in 2012 – a futuristic sky crane (above) gently lowered the rover (below) to the surface. Artist's impression.

Left: *MRO* snapped this view of Mars's largest moon, Phobos, in 2008. The large depression on the right is Stickney crater.

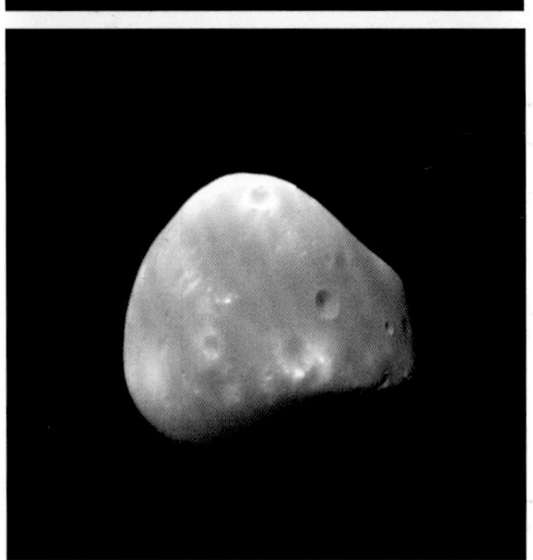

Left: Image of Mars's smaller moon, Deimos, taken by *MRO* in 2009.

Below: *Curiosity* looks back at its tracks across the surface of Mars (2014). **Inset:** Martian meteorite ALH84001.

Above: Mosaics of Mars showing Valles Marineris, the most extensive canyon system in the Solar System (left), and the notable Schiaparelli crater (right). *Viking* Orbiter data.

Below: NASA/ESA Hubble Space Telescope views of Mars at opposition (aligned Sun-Earth-Mars) between 1995 and 2007. Mars's apparent size changes depending on its distance.

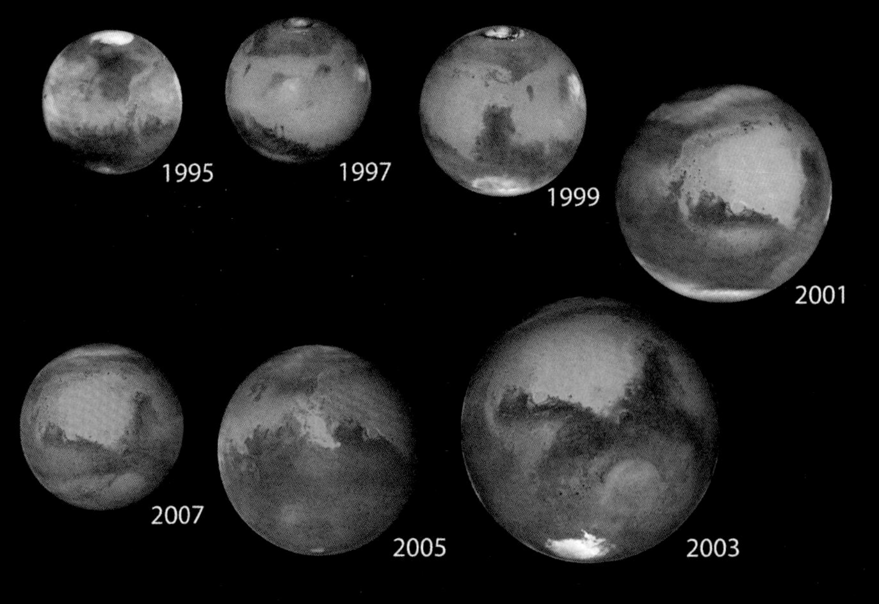

Left: Mars's southern polar ice cap as seen by *Mars Express*.

Below: Mosaic showing Syrtis Major, one of the most prominent dark patches – 'albedo features' – on the planet's surface. *Viking* Orbiter data.

Bottom: *MGS* view of Mars's northern polar ice cap.

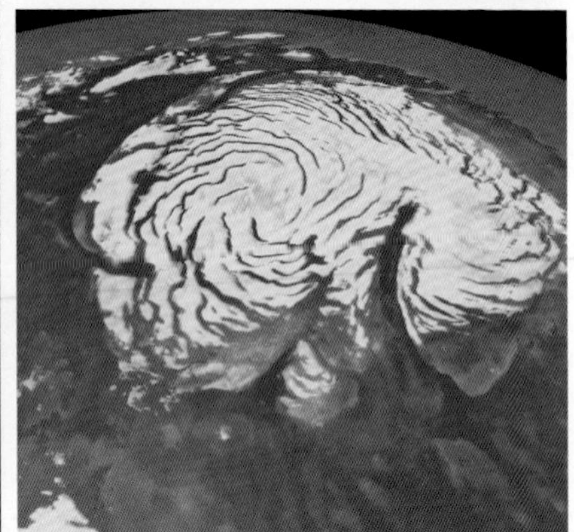

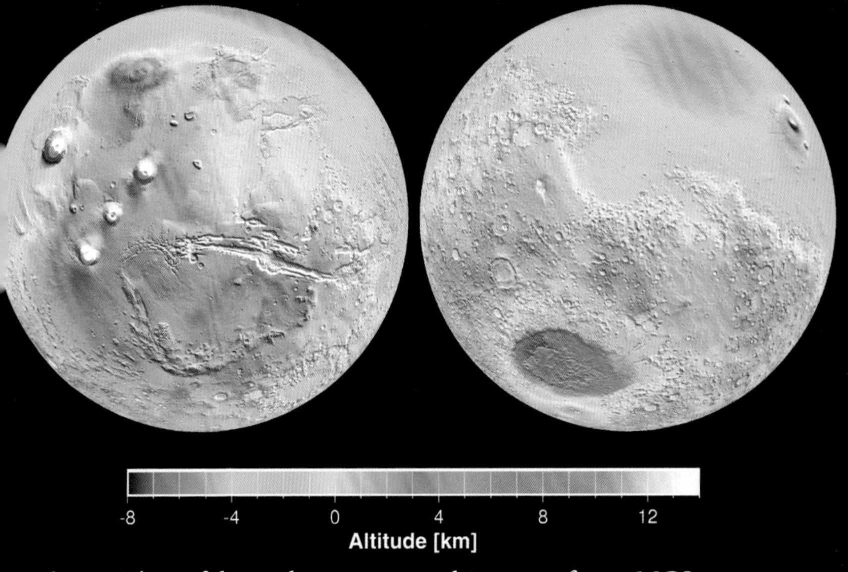

Above: These false-colour topographic maps from *MGS* illustrate the stark altitude difference between Mars's southern (red-orange-yellow) and northern (blue-green) regions.

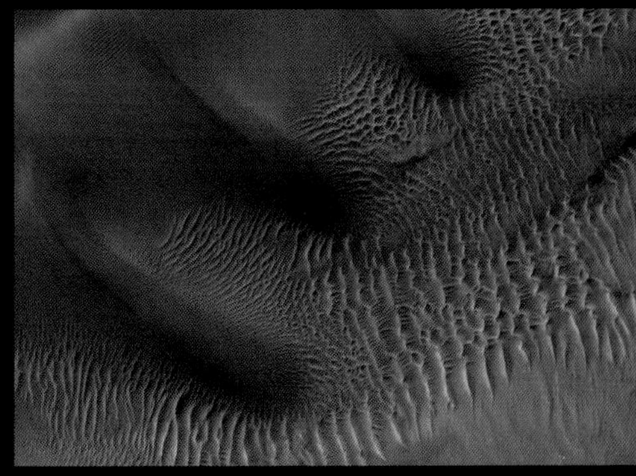

Right: Rippling sand dunes on Mars as seen by *MRO* (enhanced colour), roughly 1km (0.6 miles) across.

Right: Sunsets on Mars appear blue-tinted due to the fine dust particles thrown into the air, which allow blue light to filter through more effectively than light at longer wavelengths. *Curiosity* snapped this sunset in April 2015.

Above: Mars has experienced three major stages of life: the Noachian, Hesperian, and Amazonian periods (from oldest to present day). Each era also lends its name to a patch of Mars's surface thought to have formed at that time. Above is a slice of northern Noachis Terra, the region of Mars associated with the planet's earliest days (*Mars Express*).

Right: This butterfly-shaped crater sits within Hesperia Planum (*Mars Express*).

Right: The smooth area of Amazonis Planitia (*Viking 1* mosaic).

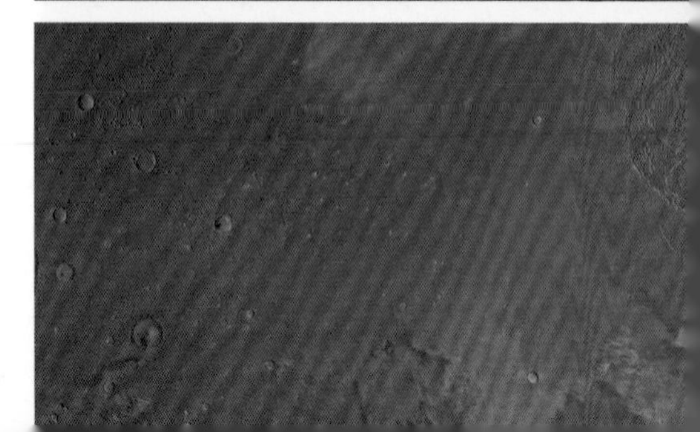

actually be a combination of *two* types of face rolled into one: that of a human and a lion placed next to one another, split down the middle. 'We have maintained for over nine years … that the eastern half of the Face, the portion that until now had only been seen in shadow, was feline in appearance,' wrote Hoagland on his website in 2001, after the new *MGS* image was published. 'The western half, we have always maintained, was humanoid. We do not believe that this object was ever intended to be a fully human visage.'

Another theory expanded on this, stating that if the left side of the Face were mirrored across its vertical axis, the mound actually resembled a pregnant human torso, with bare breasts and a pair of arms gently cradling a swollen belly. If mirrored the other way round, the right side formed a complete leonine (Lion) head. Placing the leonine head on the neck of the pregnant torso melds the forms of Virgo and Leo, as with a traditional Greek sphinx.

Other theories abound. As noted by astronomer and science communicator Phil Plait, something that's often oddly overlooked and buried in technicality is, well, the actual appearance of the mound itself. Even if you were to say that any of the images looked vaguely face- or Sphinx-like, even in the lowest-resolution frame they clearly do not look as if an intelligent civilisation consciously attempted to create a representation of a human face or form. 'Had we had the better-resolution images first, no one would have given that hill a second look,' wrote Plait on his old Bad Astronomy website. 'Hoagland's arguments don't work backwards. If it *doesn't* look like a face in a higher-resolution early image and *does* in a lower-resolution later image, then he has no leg to stand on.'

On 22 July 2006, ESA's *Mars Express* spacecraft also imaged the Face, showing the mesa with a ground resolution of approximately 13.7m per pixel (the probe has since gone on to map much of Mars at a resolution of around 20m per pixel). This image confirmed the previous consensus that the Face is a completely unremarkable, naturally occurring landform.

However, this is far from the end of the story. Since the Face was discovered, many more faces and bizarre shapes have apparently popped up all over the place, all of which can be attributed to optical illusion, tricks of perspective and low image resolution. Many of these again exist within the Cydonia region and were identified within the same 1976 batch of Viking images. Key features include the City, the Fortress, the Cliff, the Tholus and the D&M Pyramid, all of which exist in close proximity to the Face.

The City lies just tens of kilometres west/south-west of the Face. In the original Viking image – and even more so in the higher-resolution images released since – it resembles a pile of jagged rocks and uneven mounds clumped together. However, some instead saw the remains of a sprawling city, one that had once housed a vibrant and intelligent community of Martians, but that now lay abandoned. They marked the location of a plaza, a terrace, a city centre, six or seven distinctively shaped pyramids and the relics of an old fortress. They claimed that several of the structures looked unfinished, as if the civilisation were unexpectedly interrupted during the city's construction.

There's also the D&M Pyramid, which Hoagland dubbed 'the Mathematical Rosetta Stone of Cydonia'. This five-sided pyramid is named after its 'discoverers', the aforementioned Vincent DiPietro ('D') and Gregory Molenaar ('M'), who were involved in developing the Cydonian hypothesis. According to the duo, its characteristics – its sides and angles and inclines and exact positioning relative to the other structures – are apparently meaningful, unique, and very unlikely to occur randomly or naturally.

The other features are all scattered in the general vicinity of the City, Face and D&M Pyramid. The Cliff lies to the east of the Face. Rather than being a naturally occurring cliff as the name suggests, it is instead described by Hoagland and co. as a long, straight and artificial wall sitting atop a Face-like mesa. The Tholus, a small, domed, hill-like mound sitting to the south-east of the Cliff, apparently shows signs of a spiralling road snaking from the bottom to the top. The Fortress may apparently be the ruins of a fortress, as the name suggests, or the site of a collapsed pyramid (like the D&M).

Hoagland has been pretty outspoken about how these structures were intentionally carved out of Cydonian rock by intelligent aliens. He even went one step further, claiming that the Cydonia region is laid out so carefully and consciously that its various 'anomalous features' are positioned according to a geometric pattern that is undeniably deliberate. That is, if you draw lines connecting the Face, the City, the pyramids and so on, the ratios of the line lengths and angles created represent fundamental mathematical constants or relationships involving e, pi, square roots of small integers and trigonometric functions. The same is true for the sides and angles of the D&M Pyramid, he said, and claimed that it'd be impossible for this to be coincidence – it requires intelligence. He named this the 'Cydonia Geometric Relationship Model'. (In reality, this is likely due to a combination of low image quality and the deliberate selection of the lines and angles used.)

Some have mapped the same geometric model on to various cities on Earth, and found the same mathematical relationships to hold true in locations such as Washington DC in the US, and Avebury, a small village in south-west England near Stonehenge that hosts a series of Neolithic stone circles. Others drew comparisons to pre-Columbian Mesoamerica, questioning whether there may be an ancient link between a once-thriving Martian civilisation and our own. If Hoagland and co. somehow happened to be correct, our human heritage and history may be very different to what we currently imagine.

Here, it's worth referring back to Brandenberg, one of the strong advocates for the Face having been carved by a humanoid race of Cydonians. This may seem far-fetched now, but in essence the Cydonian hypothesis was just that – a hypothesis – and could potentially have been true. Some parts of the hypothesis had a reasonably solid scientific basis, for example, that Mars was once warmer, had more oxygen in its denser atmosphere and hosted an abundance of liquid water. In essence, it was once a far more habitable world than it is today. There's also an astronomical argument (dubbed the 'mediocrity principle') to suggest that there's nothing special or unusual about life on Earth, so why shouldn't similar life have once arisen on Mars, too?

In 2011, Brandenberg proposed that Mars had once hosted natural nuclear reactors, essentially huge, naturally occurring deposits of uranium ore that react with groundwater, causing nuclear reactions that self-sustain over time. We've discovered a small number of such reactors here on Earth – all in one town on the western coast of Central Africa (Oklo, Gabon) and all inactive for at least a billion years – by analysing the levels of uranium present in rock samples. Brandenberg argued that the isotope ratios in Mars's soil and atmosphere suggested that large-scale nuclear activity had once taken place on the surface of Mars. He proposed that such a natural reactor, far larger than those of Earth, had once sat in the middle of Mars's northern regions, churning out uranium-233 (a type used in nuclear weapons and as fuel in manmade reactors) until it eventually and explosively self-destructed, 'ejecting large amounts of radioactive material over Mars's surface'.

There was nothing particularly weird about this suggestion. It was based on solid isotope markers and, as Brandenberg put it, 'natural nuclear reactors formed and operated on Earth; there is no reason this could not have happened on Mars. Conditions on Mars – lack of plate tectonics and nearness to the asteroid belt – may have favoured such occurrences in larger size and duration than on Earth'. However, Brandenberg then went further and began to firmly leave the realm of sensibility.

Rather than a natural nuclear reactor, Brandenberg began to believe that this nuclear activity had actually been in the form of 'two large thermonuclear explosions on Mars in the past … near Cydonia Mensa and Utopia Planum, both locations of possible archaeological artefacts'. He claimed that certain isotope ratios in Mars's atmosphere were more reminiscent of the nuclear testing we've previously done on Earth than of naturally occurring nuclear reactions. In his words, 'the explosions appear due to very large fusion-fission devices of similar design as seen on Earth … the largest approximately 80m in radius' (260ft). He backed this up using the Face as 'evidence', noting that recent photographs – bear in mind that Brandenberg claimed this in 2014 – 'confirm eyes, nose, mouth, helmet structure, with additional detail of nostrils and helmet ornaments being clearly seen'. He noted that the images confirmed the structure of the D&M Pyramid, also showing 'collapsed brickwork', and even referred to a recently found second 'face' located at a different site on the Martian surface (Utopia).

His conclusion? 'Taken together, the evidence suggests that Mars was the locale of a planetary nuclear massacre … the Cydonian hypothesis has been confirmed.'

According to Brandenberg, the eroded remains found throughout both Cydonia and, in smaller part, Utopia, comprise archaeological remnants from a now-extinct indigenous humanoid civilisation of Martians. A 'massive nuclear attack', a 'nuclear calamity', may have ended life on Mars forever, changing the once-habitable planet into the cold, arid, ostensibly lifeless husk of a planet we see today. This could have been in the form of a malicious attack from hostile cosmic neighbours, said Brandenburg, 'from things as alien as Artificial Intelligence "with a grudge" against flesh and blood, as in the movie *Terminator* [sic], to things as sadly familiar to us as a mindless humanoid bureaucrat like Governor Tarkin in *Star Wars*, eager to destroy Alderaan as an example to other worlds.'

The power of pareidolia

Although the illusion has been long debunked in the scientific community – opponents such as Hoagland and Brandenberg may be vocal, but their views don't hold much water – the Face is still fascinating. There's no denying that it did bear at least a passing resemblance to a face in the original Viking image. Why did it look so deceptively familiar?

While much can be attributed to a mix of resolution, lighting and coincidence, a phenomenon known as pareidolia is to blame. This is when we perceive something vague or random as meaningful; we see familiar faces, patterns, shapes, or hear familiar sounds or messages, where there simply are none.

Alongside images of Jesus found on curtains and grilled cheese sandwiches – Sagan reportedly likened the Face to seeing Jesus Christ on a tortilla chip – we experience this often with the Moon and its various dark and light patches. Some see a man in the Moon, while other cultures instead see a rabbit, a female figure, a frog or a buffalo. However, the cratered and pitted surface of our moon doesn't actually show any of these shapes; our eyes are adept at finding something in nothing. You're not objectively seeing a man or a rabbit or a frog: you're looking for symbolism that's familiar to you and your culture, and fitting this to your surroundings.

While they were forced to recognise its potential influence, Hoagland, Brandenberg and co. were dismissive of pareidolia. 'There is also another well-known human phenomena: "denial",' Brandenberg wrote (in his 2014 paper describing a nuclear genocidal apocalypse on Mars that caused the death of an advanced humanoid species that spent their time carving curiously and improbably familiar Sphinx-like structures and pyramids from rock, leaving behind no other signs of their existence whatsoever), labelling it 'negligence' to 'ignore possible signs of [intelligent] activity in an unknown and dangerous cosmos'.

In the case of the original Viking images, image resolution is key. If you were to take a poor-quality image from the very first camera you ever owned and enlarge it to be a few metres across, you'd find the clarity greatly decreased and it'd seem far more pixelated. Bigger does not equal clearer.

Pareidolia is a powerful thing, and it's still in full force where Mars is concerned. Mars 'enthusiasts' continually identify many more weird and wonderful objects in each new batch of pictures, all of which are curiously – suspiciously? – Earth-like and familiar.

People have claimed to have found formations on Mars resembling a screaming man, several humanoid and hybrid 'alien' skulls, smiling faces, George Washington, US Senator

Ted Kennedy, a panda, an anteater, a gopher, a dried jellyfish, an Egyptian statue, a crocodile, a fossilised frog, dinosaur eggs and fossils, humanoid aliens casually resting on rocks, pieces of pottery, turtle remains, a human thighbone, a wrench, Kermit the Frog, a lizard, a rat, a jam-filled doughnut (oddly specific), a pitchfork, a squirrel, multiple crouched figures, a decapitated head, a tank, alien corpses, a waving alien claw, large power plants, skyscrapers and a foot from a statue with the toes missing (yes, really) – to name just a few. Who knew Mars housed such a bizarre and diverse array of Earth-like forms?

Some have also apparently spotted the head of former US president Barack Obama in photographs from NASA's *Spirit* rover – or perhaps even his entire body. 'From the way it's standing up, I would say that it's a full statue, but is buried from the shoulders down,' wrote Scott Waring on the 'UFO Sightings Daily' website in 2014, putting its 'chances of being a real sculpture head' at a whopping 98 per cent. 'I highly doubt this is mere rock; it's probably a 3D-printed substance with tech printed inside,' he added.

Odd, but perhaps unsurprising given the fact that Obama has reportedly spent some time on the Red Planet in person. This claim was perpetuated by Andrew Basiago, an American lawyer better known as an all-round voice of oddity and eccentricity in the conspiracy community and ironically named 'truth movement'. Basiago is a self-proclaimed 'indigo child' (child with supernatural abilities), 'leading scholar' and 'twenty-first-century visionary'. He also entered the US presidential race in 2016! If he had won, he intended to 'shun the mainstream media as incompetent' – after all, 'not a single member of the mainstream media sought to interview [him]' after he revealed in 2011 that 'Obama was a participant in the Mars jump room program', a kind of cosmic teleportation room used to travel from Earth to Mars. 'True history was lost,' he mourns.

According to Basiago, Obama (under the alias of Barry Soetoro) was drafted into the government's top-secret time travel programme as a teen and trained to be a chrononaut (time traveller). He visited the surface of the Red Planet at least twice in the 1980s, said Basiago, and his various tasks included making any inhabitants of Mars, animals and humanoids alike, comfortable with the presence of humans to prepare for eventual colonisation.

Unsurprisingly, the White House have denied this claim without much stress. According to *Wired* magazine, a spokesperson for the National Security Council stated that Obama spent no time at all on Mars – 'unless you count watching Marvin the Martian'.

CHAPTER NINE

Life on Mars?

In 1984, a group of meteorite hunters came across a dull, dark, blackish chunk of knobbly rock in Allan Hills, Antarctica. Weighing in at just under 2kg (4.4lbs), the rock – named ALH84001 – was later found to be far more exciting than it first appeared.

Firstly, it was a meteorite, a real-life rock from outer space. It was also truly ancient. At over four billion years old, the rock dated back to the early days of the Solar System.

ALH84001 was soon found to have a peculiar composition. It was initially classified as a rare type of meteorite known as a diogenite, a type composed of igneous rock and thought to have broken away from one of the largest bodies in the asteroid belt (Vesta). However, over a decade later, it was re-examined and reclassified as a member of the so-called 'SNC' meteorite group (short for the Shergottite, Nakhlite and Chassignite classes, each named after the discovery site of such a rock: Shergotty in India, Nakhla in Egypt and Chassigny in France)*.

At first, SNC meteorites were simply thought to be another, albeit unusual, class of meteorite – perhaps chunks of debris that had broken away from some larger rock sitting elsewhere in the Solar System, or a visitor from the asteroid belt spanning the gap between Mars and Jupiter (or a mix of

* ALH84001 doesn't fit in perfectly with its SNC cousins. The SNC classes each have distinct compositions that differ from that of ALH84001, which is instead technically something known as an orthopyroxenite (an igneous rock stuffed full of pyroxene minerals). A few other odd rocks – basalts and breccias – have been linked to Mars, so rather than rename the 'SNCs' to encompass these, scientists instead commonly refer to SNC meteorites as 'Martian' or 'Mars' meteorites.

both). However, scientists slowly realised that they may, in fact, hail from Mars: the mix of gases locked up within them was curiously similar to what we knew of the Martian atmosphere, and their compositions seemed to resemble the soil samples studied by NASA's Viking landers. In fact, for SNC meteorites to have an origin other than Mars, their parent body would need to be pretty much identical to the Red Planet. ALH84001 was a nomad from Mars!

Scientists investigated the meteorite and confirmed a Martian origin, concluding that another body must have smashed into the surface of Mars many millions of years ago, breaking off a chunk of rock and flinging it out into space. After gliding through space for many millions of years a fragment of this rock eventually touched down in Antarctica, where it sat quietly for thousands of years more before it was discovered.

Knowing that ALH84001 formed on Mars, the meteorite's age became even more exciting. Martian meteorites are rare. Of the 40,000 or so meteorites we've discovered, fewer than 200 appear to be Martian and ALH84001 is one of the oldest of these. Existing initially as ancient magma on a turbulent young Mars, perhaps spewed out as lava by an early volcano located in the Tharsis region, the rock is thought to have crystallised as Mars rapidly cooled down. The rock has experienced most of Mars's life history, reaching right back to the planet's warmer, watery and volcanic adolescence.

All of this is exciting enough, but the meteorite was more remarkable still. When analysed more closely, it was found to contain splotchy grains of carbonate, suggesting that the rock had once been submerged in water. This was remarkable at the time – and, to an extent, still is – and added to our growing excitement over Mars's early days. While we suspected that water had once existed on Mars from its surface geology, first seen by NASA's Mariner and Viking missions in the 1960s and 1970s, we had no direct evidence of that until years later. Other suspected Martian meteorites didn't show this kind of structure as they were far younger and thus presumably formed when Mars was drier. At the time, ALH84001 was the only record of a wet Mars we'd ever found.

However, the meteorite's remarkable age and origin were quickly trumped by yet another unusual characteristic. ALH84001 contained something so significant and unique that it quickly became one of the most famous pieces of rock in the world. In the summer of 1996, scientists published a paper in the journal *Science* in which they announced something remarkable: they had discovered evidence of life within ALH84001.

'NASA has made a startling discovery that points to the possibility that a primitive form of microscopic life may have existed on Mars more than three billion years ago,' said the accompanying statement from then-NASA administrator Daniel Goldin, which tried remarkably hard to sound simultaneously excited and understated. 'The evidence is exciting, even compelling, but not conclusive. I want everyone to understand that we are not talking about "little green men". These are extremely small, single-cell structures that somewhat resemble bacteria on Earth.'

Buried in the cracks between different nodules of carbonate in ALH84001 were numerous wiggly, squiggly, rope-like, tubular and sausage-like shapes reminiscent of dead and fossilised bacteria, measuring just tens of nanometres across. Many types of bacteria found here on Earth look remarkably similar, with many of the same characteristics, and have also been found within pieces of rock.

ALH84001 also appeared to show signs of bacterial activity. Tiny grains of various minerals (magnetite, pyrrhotite, greigite) were found alongside the worm-like structures. Minerals similar to these are produced by terrestrial bacteria, either as waste products or for other biological purposes. The rock also contained organic molecules that we know to sometimes form when bacteria decompose, known as polycyclic aromatic hydrocarbons (PAHs). On Earth, we find fossilised PAH molecules buried in ancient rock layers and stores of coal and oil, where they likely formed from the decomposition of early biological sources such as plants and plankton. All in all, the rock appeared to contain multiple strong signs of biological activity.

'None of our observations is in itself conclusive for the existence of past life,' explained the scientists cautiously, headed by lead author David McKay of NASA's Lyndon B. Johnson Space Center in Houston, Texas, US. 'Although there are alternative explanations for each of these phenomena taken individually, when considered collectively, we conclude that they are evidence for primitive life on early Mars.'

However, such a momentous finding – possibly the most important ever made – was not to be taken lightly and scientists immediately began studies of their own. In the following months, non-biological origins were suggested and tested for all of the phenomena spotted in the meteorite. The carbonate crystals could have formed incredibly quickly in the scorching hot temperatures triggered by impacts on Mars, or been deposited by water. The bacteria-like shapes themselves could have formed from purely inorganic processes, and, if we're taking Earth's history as an example of how life develops, we'd expect the 'bacteria' within ALH84001 to be hundreds of times larger than they apparently were. They could be tricks of the eye or laboratory artefacts, as can easily occur when peering through a microscope or preparing a sample for study. They could even be from Earth! After all, the meteorite had been here for thousands of years already.

All of the substances that McKay and colleagues found could also easily be produced by non-biological processes. In fact, scientists – including McKay's own brother Gordon, who was apparently an outspoken critic of his brother's hypothesis and had an office just 'down the hall from his brother's' – tried to cook up the same kind of mix of minerals and hydrocarbons found in ALH84001 in a lab here on Earth … and succeeded.

In the years following the 1996 announcement, scientific opinion strongly skewed towards the idea that the structures within the rock were, unfortunately, not evidence of primitive life. 'We certainly have not convinced the community,' McKay admitted to the Associated Press in 2006, 'and that's been a little bit disappointing.'

Squiggly shapes aside, ALH84001 sparked a fresh wave of excitement about life on Mars. Our analysis of the rock was certainly scientifically useful, and gave us an idea of where might be a good place to hunt for an exciting rock or soil sample from the Red Planet. Studies suggest that ALH84001 may have originated in either the Tharsis or Elysium regions of Mars, both of which are highly volcanic areas, or an eastern part of Valles Marineris named Eos Chasma, an elongated groove in Mars's surface composed of old volcanic lava and windswept material.

Valles Marineris is essentially Mars's super-sized version of the Grand Canyon. It is the largest known canyon system in the Solar System, stretching out for thousands of kilometres! We're not certain of how it formed, but it is thought to have been created when Mars began to rapidly cool down during the Hesperian period, causing the surface to crack. We've also found evidence of water within the canyon: ESA's *Mars Express* satellite has detected the presence of water-bearing and clay minerals in some areas, and other studies have spotted signs of water erosion and significant flooding. Landslides and various forms of water – small but steady flows, groundwater, sudden floods, standing lakes – are thought to have reshaped and widened the canyon over time, sweeping away sediment from the valley floor and creating branching cracks that we still see today.

Even if ALH84001 wasn't harbouring fossilised life, it formed in a watery, wet world, which was a striking finding in itself. Our view of Mars has certainly rollercoastered over the past hundred years, especially where life and water are concerned.

Lowell and colleagues in the late 1800s envisaged a thriving civilisation of Martians, an idea so captivating that life on Mars was thought to be a certainty. However, the Mariner missions of the 1960s killed this vision, instead showing a dead, moon-like world wrapped in a thin blanket of atmosphere that would struggle to support even the simplest of life forms. *Mariner 9* (launched in 1971) spied a different world again; the probe saw evidence of varied geology and past water – volcanoes, canyons, valleys, river channels – and reassessed Mars as a world that could have potentially hosted life. The Vikings (launched in 1975) agreed that Mars had once been a watery world, but shot down the idea of life after their direct searches for life and organic matter came up short, again putting Mars in the category of dead, arid and lifeless. Subsequent missions agreed that life was likely no longer present on Mars (if it ever was), but disagreed about the absence of organics and added to the growing body of evidence that the Red Planet was once a world covered in water. There are still differences of opinion in our view of the Red Planet, but the general consensus is that early Mars was a far milder and wetter world than today, and that it may well have been an ideal habitat for life.

NASA's two Viking probes, launched in 1975, contained the first – and only – experiments we've sent to Mars to explicitly hunt for extant life (which is life that's, well, alive). There's a crucial difference between hunting for habitability, as NASA's *Curiosity* rover is currently doing, and hunting directly for life itself. To confirm habitability, we need to find evidence that an environment is, or once was, favourable to life, with things like pH-neutral liquid water, soil that's not too acidic or salty, a possible source of energy and the presence of the 'building blocks' of life (the chemical elements carbon, hydrogen, nitrogen, oxygen, phosphorus and sulphur, and more developed organic molecules such as the aforementioned hydrocarbons).

'We look for all the different kinds of things that would say, "this is a good or bad environment for life",' says Ashwin Vasavada, project scientist for NASA's *Mars Science Laboratory* (which transported the *Curiosity* rover to Mars). 'That's our task, because we already know that actually finding evidence of life isn't going to be an easy question for Mars. If it was obvious, we would have found it already. Hunting for habitability takes one step back and tries to answer a question we're more confident of asking and is a precursor to asking if there's life anyway – after all, if there was never a good environment for life, then there's no real need to look for it.'

At *Curiosity*'s chosen landing site, Gale Crater, we've found all of the ingredients we know life needed in order to develop on Earth billions of years ago. This makes the crater the first habitable environment confirmed to exist on any world other than Earth. While this doesn't mean that life actually exists there, or that it ever did, it means that it *could* have done. Some of the oldest evidence of ancient terrestrial biology dates back between 3.8 and 4.1 billion years; if life also happened to arise on Mars billions of years ago, it could potentially have thrived in the conditions present there at the time.

Instead of taking a similar step back and focusing on habitability, the Viking missions decided to boldly go straight for the life.

In 1976, two separate landers touched down on Mars at landing sites located some 6,500km (4,040 miles) apart. Each was equipped with four different biological experiments: one designed to hunt for organic molecules on Mars's surface, and the other three to feed any potential microbes in the Martian soil with nutrients, gas, water, light, a simulated atmosphere and so on, to see if they gave off any signs of biological activity in response (little puffs of gas, for example) after 10 to 12 days. Each experiment did so in a slightly different way and combination, and hunted for different by-products.

Three of the four experiments returned negative results; no organic molecules were detected (thanks to *Curiosity* we now know these actually do exist on Mars, but Viking missed them), and two of the by-product-seeking experiments saw no signs of biological activity. However, one did return a positive result: Viking's 'Labelled Release' (LR) experiment.

For this, each lander reached out and scooped up a bit of Martian soil. One took a scoop from underneath a rock, the other from a sun-scorched part of the surface. To these soil samples they added a small drop of a solution packed with nutrients and radioactive carbon atoms – essentially a type of radioactive food that would produce traceable by-products – and closely monitored the soil to see if any little microbes exhaled radioactive gas as they enjoyed their meal. Excitingly, this appeared to be the case! The soil sample rapidly pushed out a sigh of carbon dioxide gas that spiked high above the Martian background level, indicating that something in the soil was metabolising and breaking down the droplet of 'food'.

However, the remainder of Viking's biological suite returned negative results*, leaving the majority of the scientific community either unenthused, unconvinced or unsure of what to make of the LR result. It was eventually declared to be inconclusive. Many scientists pointed out that there were non-biological processes that could have induced such a response, and that the gas spike itself happened far faster than we'd expect if it were due to biological metabolism. The result was widely chalked up to a combination of Mars's cold, dry, radiation-bombarded surface and thin but carbon-dioxide-rich atmosphere. Such conditions can lead to the production of heat-sensitive substances known as superoxides, essentially reactive salts that contain negatively charged oxygen ions. These could have oxidised the nutrients that the Vikings added to the soil and kicked out the observed burst of gas, but been broken down in the heated control sample.

In general, NASA deemed it more important that Viking had failed to find evidence of any organic molecules on Mars – after all, without these, life couldn't exist anyway. The initial excitement died down and the majority of scientists dismissed the idea of extant Martian life, instead attributing

* Viking's Gas Exchange (GEX) and Pyrolytic Release (PR) experiments did detect bursts of gas but these were consistent with our expectations. For each soil sample scooped up from Mars's surface, the lander collected another and sterilised it with heat (to kill off any potential life). In order for an experiment to report a 'positive' result for life, it needed to detect potential biological by-products within the non-treated sample, but not in the control sample. If a burst of gas, for example, was detected in both the experimental and sterilised control samples, we could chalk it up to a non-biological process, as we know the control sample to be lifeless. For the LR experiment, the test sample produced a 'positive' response while the control showed a 'nil' result.

the result to a surprising and unusual kind of 'enigmatic chemical activity' taking place in Mars's soil.

On the other hand, the principal investigator of the LR experiment, Gilbert Levin, disagreed. In the years since Viking – and as recently as mid-2016 – he has been regularly quoted as believing the LR experiment to have been the first detection of life on Mars. In 1997, after over two decades spent studying the LR results, he declared his belief that 'the LR had, indeed, discovered living microorganisms on the Red Planet'. Many subsequent papers – both co-authored by Levin and not – have supported this view and a level of controversy still exists today.

This is especially true given the fact that *Curiosity* has proven organics to exist on Mars in recent years. One of the key strikes against the LR result being an indication of life was the failure of Viking's other biological experiments, especially the failed detection of organic (carbon-containing) molecules. However, scientists have since realised that the experiment designed to detect organic molecules was simply not sensitive enough to do so, making it potentially 'blind to low levels of organics on Mars'.

'In the Antarctic Dry Valleys and the Atacama and Libyan Deserts [on Earth], we find [an amount of carbon] which would have been undetectable by Viking,' wrote a team of scientists, led by Rafael Navarro-González of the National Autonomous University of Mexico, in 2006. This, coupled with the fact that the oxidation process (via which carbon-containing molecules turn into carbon dioxide) can have a strong negative effect on the detection of organics, led the scientists to conclude that 'the Martian surface could have several orders of magnitude more organics than the stated Viking detection limit'.

The only organics Viking detected were the chlorine compounds chloromethane and dichloromethane, both of which were chalked up to self-contamination and dismissed. However, a later discovery made by NASA's *Phoenix* mission (launched in 2007) suggests this may have been a misguided error. *Phoenix* found substances known as perchlorates, negatively charged chlorine-oxygen salts, to exist in Martian soil. Perchlorates can sit alongside organic molecules happily and unobtrusively, but as soon as they are heated – as Viking's organic-molecule-hunter did – they aggressively break down any organics in their vicinity. By heating the soil to hunt for organics, Viking may have actually been destroying them, making a negative result inevitable. Researchers tried mimicking this using sand from Chile's Atacama Desert, an Earth environment commonly used as an analogue for Martian terrain. They scooped up soil they knew to contain organics, added perchlorates, analysed it as Viking did – and detected chloromethane and dichloromethane!

Regardless of whether it signals the presence of life on Mars, the process involved in the LR mechanism is not at all well understood and future missions to the Red Planet will have to explore further. Overall, it seemed that Viking had touched down on a dead and uninhabited world. This was devastating for many and had a big impact on our drive to explore Mars. Following the Viking duo, NASA's next attempted launch to Mars was in 1992 (and the next successful mission took place in 1996, two decades later).

Speaking to the *Washington Post* in June of 2016, physicist and former astronaut and NASA chief scientist John Grunsfeld attributed this substantial gap to Viking's desolate view of Mars. The probes painted a picture of an apparently lifeless Mars – 'I think for 20 years that put the damper on the Mars program,' said Grunsfeld.

Extant, extinct, extreme

Although we know that Mars was once at least partially habitable, we have found no signs of life on the planet. There is a real spectrum of opinions floating around in the scientific sphere and the general consensus is somewhat cautious; while it's possible that life existed on Mars, we simply don't have enough evidence to confirm or deny the existence of Martian life (either past or present).

'In fact, I'd say that there's a three-axis debate,' says John Mustard, professor of Earth, Environmental, and Planetary Sciences and Environmental Studies at Brown University, Rhode Island, US. 'Each axis has passionate supporters who have influenced the debate about our search for life on Mars.'

Two of these axes hit the extremes: that life either currently exists on Mars – 'fuelled, in my opinion, by enthusiasm unconstrained by data', says Mustard – or that life never evolved there at all. The third option is that of extinct life, life that once existed but has since died off. 'This is fuelled by optimism that the ingredients for life – water, energy, nutrients – were present, something that is substantiated by data and observation,' says Mustard. 'The optimism is that some form of self-replicating life would spontaneously emerge, and that Earth cannot harbour the sole genesis.'

The idea that life never developed on Mars is the simplest to summarise. Mars's history has been remarkably changeable and the planet today is far from friendly. Many things can stop a planet being habitable: the presence of an atmosphere and magnetic field, temperature, seasonal variations, presence and abundance of water, chemistry, availability of suitable energy sources, sheer bad luck when it comes to incoming space rocks and, importantly, location. Earth's distance from the Sun places it smack-bang in the middle of our star's 'habitable' or 'Goldilocks zone', defined as the region within which liquid water can exist on a planet's surface (assuming the planet has enough atmospheric pressure to support it). In most estimates, Mars sits just slightly too far from the Sun to be included*.

Despite our long-term monitoring of the planet we have not found any signs of life on Mars (nor on any other world

* There are other estimates for the Sun's habitable zone that do include Mars's orbit, but the planet still fails to fulfil the caveat of having sufficient atmospheric pressure to maintain stable liquid water on its surface (this is only thought to be possible in anomalous and short-lived situations, if at all).

in the Solar System). Although many believe that life is likely to have emerged elsewhere in the Universe, we have not found any evidence that this is the case. It's very possible that life simply never took hold on Mars in the first place.

The idea of extant life on Mars is somewhat more exciting. Could some kind of simple organism – or even something more developed – currently exist there?

The general view is that it is unlikely, but possible. Multiple spacecraft have monitored Mars constantly for years, we've photographed and topographically mapped the entire surface, and directly scooped up its soil and drilled into its rock. Despite Viking's ambiguous results we've discovered no signs of life, and even the most alien of life forms would push out by-products and show regular and detectable signs of biological activity.

From this, all we can say is that life almost certainly doesn't live on the surface of Mars. Indeed, the surface conditions today appear to be pretty hostile: temperatures are low, water cannot survive in a liquid form, and the levels of incident radiation (from both the Sun and the wider cosmos) are fierce and damaging. If we were to find extant life on Mars's surface it would need to have adapted to handle high levels of radiation, extreme aridity and very low – and wildly variable – temperatures.

However, there's a chance that simple microbial life could exist underground or in localised groups where a specific habitat is particularly favourable. It might be locked up within ice sheets and groundwater flows, or perhaps within canyon crevices or overhangs, cracks, fissures or underground lava tubes. Essentially, it might exist in pockets within cave-like environments.

Life needs very different conditions to survive and adapt than it does to emerge or flourish. Life may have developed on the surface of Mars when it was far more habitable and moved underground as the planet's climate changed, adapting or lying dormant and saving itself in the process. It could therefore exist in deeply buried material, which could perhaps be found on or beneath the floors of impact basins (which naturally burrow deeper into Mars and dig up older regolith), or in rocks that have been flung around and ended up somewhere more temperate, such as a hot spring or lake.

This is not just speculation. We believe there to be many underground caves – perhaps excavated by or linked to volcanic activity – lying beneath Mars's surface. If Noachian or Hesperian Mars's internal heat supported ancient aquifers or springs (paleosprings) at these underground locations, as we believe to have been the case, life may well be thriving there happily undisturbed.

'If you were *me* looking, you would clearly look underground. I believe that extant life could still exist at depth at places on the planet where the geothermal heat flow from the planet's interior might intersect with the presence of water ice, providing a type of habitat that we have been studying here on Earth,' says Penelope Boston, director of the NASA Astrobiology Institute in California, US. 'I think it is a valid and tantalising suggestion that has become steadily more plausible as we have learned more about subsurface occurrences of water ice, the new results confirming leaking fluids at the RSL features*, and other complexities of even the near-subsurface chemistry, which appears to be quite different immediately below the surface.'

If life were to exist in subsurface conditions such as this, our current technology deployed at Mars is simply unable to detect it. To either find life or discount the possibility we would need to drill deep down into Mars's regolith, which is no mean feat. 'Drilling to depth is a great technological challenge,' adds Boston. 'The possibility of extant life cannot be ruled out without extensive exploration over the course of many upcoming missions, including, eventually, possible human missions.'

---

* RSLs are dark streaks formed by seasonally flowing salty brines. Images from NASA's *Mars Reconnaissance Orbiter* have shown RSLs flowing down various crater walls on Mars.

We have followed a simple game plan when hunting for life on Mars so far – we follow the water*.

'For the Mars science community, there are two reasons to think you should look for water as a starting point,' says Ashwin Vasavada. 'One is just the fact that it's so important for life on Earth – and ubiquitously so. There are a lot of different forms of life on Earth, and water is the common denominator. We don't know of any form of life that doesn't use it. And then, from a theoretical abstract, if you're really going to think about life in the Solar System or Universe apart from Earth then you just look at the chemical properties of different materials. It's no coincidence that life on Earth uses water.'

This is a mix of 'go with what you know' and of savvy science. Even if we set aside the fact that terrestrial life relies on it, water is still a relatively unique substance. For one, it is an especially good solvent, so much so that it's been nicknamed the 'universal solvent'. This is due to its composition and structure: each water molecule comprises one oxygen and two hydrogen atoms bonded to form a bent 'V' shape, with the oxygen atom sitting at the central point and the two hydrogen atoms leaning away. Both elements act differently around electrons, with oxygen attracting more electrons than hydrogen does. Electrons thus clump around one side of the water molecule and leave it slightly negatively charged (oxygen), while the other side skews slightly positive (hydrogen).

Because of this polarity, other polar molecules and ionic compounds (such as sodium chloride [salt], where each ion is charged and thus differentially attracted by water's polarity) dissolve readily in water, while non-polar molecules clump together and stay separate (oils, fats). This latter effect, known as the hydrophobic effect, is crucial in forming and maintaining cell membranes.

---

* Our approach is so water-centric that NASA's official strategy in their hunt for extraterrestrial life (on Mars and elsewhere) is literally 'Follow the Water'.

Also, at risk of stating the obvious, water flows. It can move substances around and transport all kinds of nutrients, minerals and chemicals towards a cell, or waste products in the opposite direction – something that's crucial for life, especially when coupled with water's admirable properties as a solvent. It also morphs from solid to liquid to gas in a relatively small temperature range, with the liquid phase existing at the perfect point for Earth, and as such can create a diverse range of habitats for life. There are far more properties that make water special: its transparency, conductivity, density and more. Other substances (methane, ethane, ammonia) can do similar things to water, but they're not good all-rounders. We haven't yet confirmed the existence of any large liquid stores of ammonia in the Solar System, but we know that large stores of methane and ethane exist on Saturn's moon Titan – an exciting prospect.

As a result, anywhere on Earth there's water, life has somehow evolved a way to make use of it. This could also have occurred on early (wet) Mars.

Martian life would have needed some kind of solvent in order to develop. We know that Mars once had water – and lots of it. It makes sense that potential early life may have used water as a solvent due to its availability and suitability. Early Mars had various energy sources (volcanic, sunlight), organic material, was more protected from cosmic radiation and was overall a far more temperate place. The availability of water may have been the only real constraint on the emergence of life (an event dubbed abiogenesis) on Mars, hence why we focus on places we know that water once existed (old riverbeds, gullies, wherever there are signs of water erosion) and get excited about locations in which it might still exist today (underground lakes, polar caps, melting groundwater and brines such as the RSLs).

When considering the watery nature of early Mars, life wouldn't need to have been especially resilient or bizarre in order to grow and thrive. However, given the Red Planet's turbulent and variable past, it likely would have needed to think fast on its little alien feet in order to survive. For example, Mars's quickly thinning atmosphere caused the planet's liquid water to rapidly leave its surface, forcing it both upwards and downwards, meaning that any microbes happily paddling about would have needed to somehow find another wet home. While it's highly possible that life didn't, or couldn't, adapt or relocate quickly enough to survive, it may be more nuanced than that.

There are some exceptions to this mass exodus of water. Bodies of water that could somehow maintain their own heat – likely by some kind of volcanic heating or hydrothermal system, as with the hot vents we see on Earth – would have lasted far longer and would possibly still be somewhat habitable.

This is something that is key when comparing Earth to Mars; Martian life would have had to be incredibly hardy to endure the stop-start conditions it would have experienced on ancient Mars. Life would have had to develop very differently to that on Earth because things kept being interrupted by impacts, the loss of an atmosphere it may have used to breathe, loss of water it may have been living in or feeding from, increasingly powerful radiation hitting it and breaking down its cellular structure and more. On Earth developing life had steady access to consistent water, energy (sunlight), carbon, nutrients and the like, whereas life on Mars would have had to be far more adaptable or able to shut itself down and shift into 'survival mode', or have developed in different areas simultaneously.

Much of our uncertainty in this area stems from our relatively new understanding of terrestrial organisms that are startlingly resilient and even somewhat alien: extremophiles. Extremophiles (literally 'lovers of extremes') are organisms that somehow survive and even thrive in incredibly extreme and hostile parts of Earth. We started to learn more about these otherworldly life forms in the 1980s and 90s, and have since found them to be amazingly adaptable.

We've found extremophiles on Earth that have adapted to extreme temperatures, acidities, salinities, drynesses, pressures, lack of nutrients, absence of sunlight or oxygen, incredible amounts of radiation and more – all conditions that would be completely inhospitable to humans or complex life. Some, named chemoautotrophs, can even draw energy out of rock (crudely put, by transferring electrons between the electronic states present in different types of mineral). Many of the environments we've found hosting such organisms are at least partially Mars-like in one aspect or another. These extreme organisms may even show how life first developed here on Earth; some biologists have suggested that terrestrial abiogenesis may have occurred deep underwater around hydrothermal vents.

'Those of us who work in extreme environments are more likely to guess that life probably also arose on Mars than those who work in other parts of astrobiology,' says Boston. 'My own view is that life most probably did arise on Mars, if our current understanding of its early history as a much warmer and wetter body with a denser atmosphere proves to be true. Focusing on the subsurface as I do, I would look for organisms that can get the energy they need to live from the inorganic chemical reactions that we see here in the subsurface of Earth (namely, the transformation of minerals containing elements like manganese, iron, sulphur and other compounds).'

If life were of the extreme kind, it could look fittingly alien. Many of the extremophiles we see here on Earth are truly bizarre, with all kinds of interesting and unusual features, extremities and colourings. They also seem to be much smaller in size than 'average' bacteria on the surface of Earth. 'Why that's true we don't know, although we have conjectured that they are under different ecological pressures below ground than the guys living on the surface,' says Boston. 'Clearly, we will need to be alert to the potential for very different sizes, biochemistries, and morphological appearances. I don't know what "Martian" organisms would look like exactly, but I'm sure they would be very cute!'

Uncertainty aside, we know that Martian life wouldn't take the form of little green men or technologically advanced robot overlords. Instead it would be simple microbial life, small microorganisms like bacteria, or something akin to an algae or lichen*. This is due to the timeline on which Mars as a planet has evolved. It began warm, wet and wrapped in a thick atmosphere, but very quickly – after just a billion years or so, if that – began to morph into the dry and arid world we see today. Assuming life popped up on Mars as early as it possibly could, which in itself is quite an assumption, it still would have struggled to evolve to any level of complexity.

Comparing abiogenesis on Earth to a possible timeline for abiogenesis on Mars makes things slightly clearer. Life is thought to have appeared on Earth around 3.8 billion years ago – the planet itself formed 4.5 to 4.6 billion years ago and the Moon shortly after that.

This ancient life took the form of simple single-celled organisms named prokaryotes, which are some of the very simplest life forms imaginable. They lack organelles, which are constituent cellular 'organs' such as mitochondria (hammered by rote into many a student mind as 'the powerhouse of the cell'!), nuclei and so on. Even amoebae, common examples of simple life, are more developed! More complex life – in the form of eukaryotes, organisms whose cells have membranes and distinct organelles – took even longer still, arising roughly two billion years ago. For billions of years the Earth was only populated by microscopic amoebae and protozoa, essentially small, unicellular creatures that wriggle around and reproduce asexually by splitting themselves in half. Multicellular life finally appeared between 600 and 900 million years ago, and the Cambrian explosion (an immense but short-lived burst of evolutionary activity in which a staggering diversity of animals seems to have emerged) occurred 542 million years ago. Single-celled

---

* As with everything in this chapter – and much of this book! – it may be more accurate to say that it's *almost certainly impossible* for more advanced Martian lifeforms to exist; unwieldy caveats are still technically needed because we simply haven't enough information to discount it once and for all.

microbial life ruled the Earth for the vast majority – around three-quarters! – of its existence.

With this in mind, it's easy to see why we don't expect to find any kind of developed life on Mars. It took us literally billions of years to get to where we are today. Life on Mars simply didn't have enough time to do the same before its habitat disappeared.

This approach is admittedly quite a self-centred one. It is of course possible that Earth simply had a much tougher time in developing life, and that it all progressed far more quickly and smoothly on Mars. We may also be missing Martian 'biomarkers' because of our geocentric biases – this concern was raised when exploring the suspected biological by-products locked up within the ALH84001 meteorite. ALH84001 highlighted how difficult it might be to recognise a sign of Martian life even if it were right in front of us. For example, we assume iron oxides and hydrocarbons to be possible signs of a biological origin, but we may be missing the mark.

'Our task is difficult because we only have a small piece of rock from Mars and we are searching for Martian biomarkers on the basis of what we know about life on Earth,' wrote David McKay and scientists in the initial 1996 *Science* paper. 'If there is a Martian biomarker, we may not be able to recognise it unless it is similar to an earthly one.'

The assumption that Martian life would be carbon-based*, like terrestrial life, is also geocentric, and has been dubbed 'carbon chauvinism'. There are certainly alternatives, the most prominent of which is silicon. Silicon has a similar

---

* Carbon-based means that all of the complex molecules and chemicals needed for life – DNA, RNA, proteins, fats, carbohydrates, tissue, muscle – comprise carbon bound with other elements. Carbon can form four bonds, meaning it can latch on to four other atoms at once, but is special because its small size also enables it to form extra-strong double bonds. Carbon bonds readily and strongly with itself, and the resulting chain is sturdy and stable, making it ideal for life.

chemical structure to carbon, but differing properties. It bonds to itself more weakly than carbon does, is more reactive, and would be unable to use water as a solvent, so there'd need to be another present (methane would work, but any environment would thus require both abundant silicon and methane). It's possible that extraterrestrial life could be silicon-based, but it wouldn't be a case of simply replacing the carbon within us with silicon. Silicon-based life would be truly alien: the entire structure and system of life would need to be reimagined to suit the properties and chemistry of the silicon atom.

We also look for 'out-of-equilibrium' gases as possible markers of life. These are gases that would usually naturally degrade or combine to form another substance, but exist in a planet's atmosphere in higher-than-expected levels because they are regularly replenished by some ongoing process occurring on the planet. On Earth such gases are oxygen and methane, which are produced by plant life and bacteria respectively. Without trees pumping oxygen or bacteria puffing methane out into the atmosphere, these gases would not exist in such quantities in our atmosphere. The fact that they have not degraded or combined to form other substances signals that their levels are somehow being topped up (and in our case, we know that life is to blame).

NASA's *Curiosity* rover has detected spikes of methane on the Red Planet in past years; this excited scientists as we know methane as a biological marker. In the thin atmosphere of Mars any methane would also be subject to intense incoming radiation, so we would expect it to break down even more swiftly. We're still unsure of what provoked it on Mars, but believe it likely to be some kind of geological process or reaction, such as water interacting with Martian rock.

Despite our incomplete knowledge, we can confidently extrapolate our knowledge out into the Solar System. All life is defined and restricted by its environment. For example, we know that life on any terrestrial planet would likely not be silicon-based, as mentioned before, because huge amounts of the silicon present on such planets is trapped and locked up within the rocky crust and mantle, and is thus not available for life to use to piece itself together. We may get excited over spikes of methane in alien atmospheres, but we also know that different atmospheres would facilitate different, and predictable, out-of-equilibrium gases that we can hunt for instead.

Of hitchhikers and hardy water bears

Other scientists have far more intriguing suggestions for the origin of life. Perhaps, they say, all terrestrial life was actually alien to begin with. This theory is known as panspermia (literally 'all seeds' or 'seeds everywhere') and proposes that life came from the stars, travelling through space on a meteoroid, comet, or other space rock, perhaps originating on another planet entirely. This sounds like true bona fide science fiction, of lizards inhabiting human bodies, alien abductions and UFO sightings – but it isn't.

Firstly, space is full of complex molecules. Large interstellar clouds of hydrocarbons and organic molecules (grains comprising carbon, oxygen, nitrogen and so on) are littered throughout the cosmos. We've found that comets contain the same stuff and could potentially even carry bacterial spores, dormant bacterial 'seeds' that have essentially hibernated and switched to standby mode.

Secondly, as demonstrated by our research into extremophiles, life can be incredibly resistant. Given the diversity and hardiness of extremophiles, it is hard to propose that life is unique to Earth. We've found the aforementioned bacterial spores viably preserved in ancient materials here on Earth, including the body of a 30-million-year-old now-extinct bee trapped in amber and a 250-million-year-old salt crystal from Texas. Spores have also been tested in a simulated 'Martian' environment by mixing them into Mars-like soil – they survived – and also exposing them to the harsh deep-space environment – again, they survived and could potentially do so for years if they were shielded by rock.

We've tested other forms of life and found more of the same. Microbes have survived being stuck on the outside of the *International Space Station* for around a year and a half. Microscopic animals known as tardigrades (charmingly nicknamed 'water bears' or 'moss piglets' for their lumbering gait and usual habitats), some of the toughest organisms we know of, have held up well, too. Experiments have shown that these little moss piglets can survive in outer space for at least 10 days and likely longer. Specifics aside, scientists deem it reasonable that simple life could survive an interplanetary trip.

Thirdly, material is exchanged quite readily within the inner Solar System, extending back to the time of the Late Heavy Bombardment. During this time, both Mars and Earth were pummelled by impactors and numerous rocks from their surfaces were flung out into space. While many rocks simply fell back to re-impact their parent body, many set off into deeper space and found their way to a planetary neighbour, where they set up home. Essentially, the planets had a giant rock fight! We know that the emergence of life on Earth overlaps exactly with this violent time period, both occurring between 3.8 and 4.1 billion years ago, making the idea of transferring material a particularly exciting one. Quite a bit of material is thought to have travelled back and forth between Earth and Mars, despite the apparent scarcity of Martian meteorites.

Other signs point to the possibility of Mars in particular as being a good incubator for early life. For one, we know young Mars was likely habitable. One other sign lies in our carbon-based and water-reliant chemistry. Biopolymers – biologically occurring polymers such as DNA and RNA – actually don't deal especially well with water and find it relatively corrosive. Despite all of water's good qualities, it's hard to build genetic material in water: it is too unstable, and simply breaks into pieces.

A molecule like RNA is thought to have first formed as other complex molecules interacted with one another, but making RNA is by no means an easy or guaranteed output from such reactions. Instead, the molecules mingle and produce tarry, gunky, asphalt-like products 'better suited for paving roads than supporting Darwinian evolution', said organic chemist Steven Benner at the 2013 geochemical Goldschmidt Conference*.

According to Benner, the presence of certain compounds and minerals – borates (boron-containing chemicals) and molybdates (chemicals containing the lesser-known element molybdenum) – can interact with RNA's precursors, stabilise them, rearrange them, block them from their self-destructive tarry life choices, and catalyse them to form more of the substances needed to form RNA and, thus, self-replicating life.

This chemistry may have existed on the early Earth, but scientists think it unlikely that borates and molybdates existed in large enough quantities. Both substances also need to be in the presence of oxygen to form and remain stable, and we know that the young Earth didn't have much oxygen at all (it only came about much later, once bacteria learned to photosynthesise). Conversely, we think young Mars had plenty of oxygen; last year, *Curiosity* deduced that Mars likely once had lots of oxygen in its atmosphere when it discovered unexpectedly high levels of manganese oxide in Martian rocks, something that needs either oxygen-rich conditions or microbes to form. Additionally, the young Earth had a lot of water and assembling RNA in water is hard work. Borates similarly favour more arid conditions … such as those on Mars.

Benner has therefore proposed that drier conditions may be more suitable for forming life, rather than the idea of a

---

* Benner dubbed this 'the Tar Paradox'. Associated issues concerned the nature and formation of biopolymers themselves – namely that multiple would need to have formed simultaneously (DNA, RNA, proteins), that different structures and sizes are optimal for genetic purposes than for catalysis, and that RNA actually appears to facilitate its own destruction far more often than it should.

'primordial soup'. 'Recent data suggests that such environments might even be found today on Mars,' wrote Benner, suggesting a dry spot on the surface of early Mars, open to the air, with lots of available oxygen. Life could then have hitched a ride on a chunk of rock – maybe one similar to ALH84001 – and made its way to Earth.

This is just one school of thought and, as with any theory, there are many scientists who disagree. There may have been dry, desert-like patches on Earth that are suitable, rather than us reaching out to Mars. Benner himself expressed excitement about any outcome that could help us to learn more about our origins. As he told *The New York Times* in 2013, 'I really don't have a dog in this fight. It could go either way and I would be equally happy.'

We have a long way to go before we can confidently state that we're all actually Martians in disguise – but finding some signs of life on Mars would certainly add weight to the theory, especially if its biochemistry resembled our own.

CHAPTER TEN

Robot Cars on Mars

Mars is a planet entirely populated by robots. It is home to a mix of crashed, uncommunicative, retired and operational machines, some happily trundling around on its surface, some watching over the planet diligently from orbit and some sitting, still and lifeless, marking the position and moment they last communicated with Earth.

The planet's most famous residents are arguably NASA's three robotic rovers, two of which – *Opportunity* and *Curiosity* – are still pootling about happily today.

These rovers may have hogged the limelight in recent years, but they are relatively recent additions to the Martian population. We've been trying to reach Mars in various ways for a long time. Our very first attempts took place in 1960, nearly a decade prior to our first footsteps on the Moon and just a few years after the first satellite was sent into orbit around the Earth! Overall, Mars has been the target of more attempted launches than any body in the Solar System bar the Moon. Numerous probes have attempted to fly past, orbit and land on the Red Planet, with varying degrees of success.

Our exploration of Mars has its roots in the Space Race, an influential period of time in the 1950s when the US and Soviet Union competed to show off their technological prowess. In 1955, both countries announced their intentions to launch artificial satellites into orbit around the Earth within the next few years, a milestone that the Soviet Union hit first with the launch of *Sputnik 1* in 1957. They built on this success in 1961 by sending the first human, Yuri Gagarin, into space. Just three weeks later, the US retaliated by sending the first American – Alan Shepard, famous for playing golf on the Moon as part of the *Apollo 14* mission – into space in the *Freedom* 7 craft.

Conscious that the US needed to catch up with the Soviet Union in order to win the race, then-president John F. Kennedy considered two main options for America's space programme: the establishment of a space station in orbit around the Earth, or landing on the Moon. He settled on the latter, and in May 1961 delivered a speech to Congress in which he stated that the US 'should commit itself to achieving the goal, before this decade is out, of landing a man on the Moon and returning him safely to the Earth'. The race was on!

The following years saw various achievements – and, sadly, casualties – on both sides. The race came to a head in July 1969 when Neil Armstrong, commander of *Apollo 11*, stepped out on to the lunar Sea of Tranquillity, becoming the first human to set foot on not only the Moon, but any rock other than Earth.

While the Space Race is often thought of in terms of a desperate battle over the Moon, it also drove a number of missions destined for other targets – most prominently, Mars.

The Soviet Union made Mars a high-priority target. They launched their Soviet Mars programme in 1960, began developing a series of probes bound for Mars and wasted no time in sending off numerous volleys of missions in quick succession. The first Mars-bound spacecraft to sit on the launch pad was *Korabl 4*, also dubbed *Mars 1960A* (or *Marsnik 1* by members of the press, a portmanteau of 'Mars' and 'Sputnik'). *Mars 1960A* launched in October of 1960, followed just days later by *Korabl 5* (also *Mars 1960B* or *Marsnik 2*).

Both probes had a mass of around 650kg (1,430lbs) and were identical in design. With two solar-panel-covered wings, a cylindrical body and around 10kg (22lbs) of scientific equipment on board, the probes were designed for a Mars fly-by – a good place to start. While still tricky, fly-bys are simpler to execute than orbiter, lander or rover missions, as the portion of the mission dedicated to traversing the tricky Martian atmosphere and landing can be ignored. The Marsniks aimed to investigate the space between Earth and Mars, to see what effect interplanetary flight had on their instruments, to approach and study Mars, and to snap a few pictures as they flew past*.

Sadly, neither Marsnik made it. While they launched successfully, a fault with their thruster pumps meant that neither craft reached orbit and both fell back to Earth after reaching an altitude of just 120km (75 miles). For context, 'space' is thought to begin at an altitude of 100km (62 miles) and the US allows any pilot flying above approximately 80km (50 miles) to claim the title of 'astronaut'.

With characteristic resolve and determination, the Soviet Union decided to simply try again. A third attempt, named *Sputnik 22* (*Korabl 11* or *Mars 1962A*), came just two years later in October 1962. Designed as another fly-by mission, *Sputnik 22* was bulkier than its predecessors, with a mass of just under 900kg (1,984lbs). It got a little further than the previous two launches and successfully reached orbit around the Earth before disaster struck. *Sputnik 22* either burned up while stabilising in orbit or saw the upper section of the spacecraft (known as an 'upper stage') explode after it began to burn its thrusters and manoeuvre to a Mars-bound trajectory.

Sputnik 22 broke up into numerous smaller pieces, many of which stayed in orbit around our planet for several days. The launch took place on 24 October 1962 – probably the worst possible timing for the Soviet Union to be playing with rockets and causing mysterious explosions. That month saw the US and Soviet Union enter the Cuban Missile Crisis. The debris from the *Sputnik 22* failure was actually detected by the US and caused a spike of fear over whether it might have been part of a Soviet missile attack.

Just one week later, the Soviet Union tried its luck again with the appropriately named *Mars 1* spacecraft, followed a few days after that by *Sputnik 24* (also *Korabl 13*).

---

* Due to the secrecy surrounding Soviet space-related activities, it remains unclear whether *Korabl 4* was intended for Mars. NASA's National Space Science Data Center notes that some Soviet scientists claim no knowledge of the mission, but that others have confirmed that it was indeed a failed Mars-bound mission.

Sputnik 24 was the most ambitious mission yet. Although the Soviet Union hadn't yet mastered the fly-by, *Sputnik 24* was designed to actually land on the Martian surface. Sadly, this was a case of too much too soon; after reaching orbit around the Earth, *Sputnik 24* mimicked its predecessors and burned up when transitioning to a Mars-ward trajectory.

As with the other attempts, *Mars 1* was designed for a Mars fly-by, and aimed to explore the planet's magnetic field, hunt for belts of trapped radiation (as seen around the Earth), study the composition and structure of the Martian atmosphere, and monitor the properties of interplanetary space. *Mars 1* launched as planned, soaring into orbit around the Earth, adjusting to its assigned path and deploying its solar panels successfully. However, the launch wasn't completely free of bad news; one of the gas valves in the spacecraft's orientation system sprang a leak soon after heading into deep space. Luckily, scientists were able to swap *Mars 1* on to a different stabilisation system. Disaster averted?

Sadly not. Around five months later, the orientation problem caused a far more serious fault: *Mars 1* went quiet, losing all communication with Earth. However, it continued on its way to Mars and carried out its fly-by in June of 1963, passing within 193,000km (120,000 miles) of the planet. Technically *Mars 1* was the first craft to perform a fly-by of Mars, but seeing as all communications died prior to the fly-by it'd be pretty cheeky for the probe to claim the title. *Mars 1* continued out into space and settled into orbit around the Sun, where it remains today.

In 1964 the US entered the race for the first time, with somewhat better results. The first attempts came in the form of Mariners *3* and *4*. As their names suggest, both were part of the Mariner programme, which had already sent *Mariners 1* and *2* to explore Venus in 1962. While *Mariner 1* was aborted shortly after launch due to a software failure, *Mariner 2* was wholly successful in its fly-by mission and became the first robotic probe to encounter a planet other than our own.

Mariners *3* and *4* suffered eerily similar fates to those of their Venusian predecessors. *Mariner 3* launched successfully and headed off towards Mars, impeded only by the fact that its protective shield failed to detach, meaning that none of the instrument sensors could start working. Due to the unexpected extra weight of the shield, *Mariner 3*'s trajectory slowly changed from the projected path and it moved further and further from its intended target, eventually missing Mars entirely.

However, *Mariner 4* followed in the groundbreaking footsteps of *Mariner 2*, snagging the first ever successful – and communicative! – fly-by of Mars.

Mariner 4 was the first spacecraft to ever see Mars up close and personal, and sent back the first ever photographs of the Red Planet. The probe reached 9,846km (6,118 miles) from the planet's surface at closest approach and spied the now-famous ruddy surface in better detail than ever before. It flew over a region of Mars that was heavily cratered and resembled the surface of the Moon, with erosion features that are thought to have been caused by liquid water at some point in the planet's past. *Mariner 4*'s on-board video camera mapped roughly 1 per cent of the Martian surface in total. During its fly-by it also managed to calculate the first estimates for the planet's atmospheric pressure (lower than expected) and temperature (around -100°C/-150°F), and hunted for a magnetic field and radiation belts (both absent). *Mariner 4* happened to spy the rougher and more cratered parts of the planet, painting a slightly misleading picture of Mars as a whole. However, it shattered our previous visions of a Red Planet populated by bizarre creatures, or as a home to parched Martians desperately digging canals. The mission spied no signs of life – not just intelligent life, but life of any kind.

The US continued its Mars-bound Mariner programme with *Mariners 6* and *7*, both of which launched in early 1969. *Mariners 8* and *9* followed in 1971. Of the four, three worked as planned – during launch, *Mariner 8*'s upper stage began to oscillate and lost control, causing the spacecraft to tumble down to Earth and land in the Atlantic Ocean some 560km (350 miles) north of Puerto Rico.

Mariners 6 and *7* fared far better. The twin spacecraft skimmed 3,430km (2,130 miles) above Mars's equator and southern hemisphere, sending back around 200 pictures covering 20 per cent of the surface. These images were surprising; given *Mariner 4*'s earlier images, many were expecting to see more of a Moon-like surface, with ancient pockmarked terrain. However, *Mariners 6* and *7* revealed a far more varied and alien surface with a mixture of cratered ground and smoother plains. They also identified the south polar cap as being composed of mostly carbon dioxide ice, and obtained estimates for Mars's mass, radius, shape and atmospheric composition.

Mariner 9 became the first spacecraft to successfully orbit another planet. Its payload was similar to those carried by *Mariners 6* and *7*, but its propulsion system was far more substantial due to its more ambitious objectives, bringing the spacecraft's mass to just under 1,000kg (2,200lbs, double that of *Mariners 6* or *7*). While previous missions had snapped a few photographs as they winged their way past Mars, *Mariner 9* had the distinct advantage of staying close to the planet for just under a year (349 days). This turned out to be even more of an advantage than expected given the conditions at Mars upon the probe's arrival, as explained later.

Mariner 9 returned a huge amount of data, sending back 7,329 images covering at least 80 per cent of the Martian surface, and photographing Phobos and Deimos. Its images of Mars showed cratered terrain, as before, alongside all manner of geology: canyons, ridges, erosion features, mounds, volcanoes and more, and a variety of weather conditions, from global dust storms through to strong winds and fog. It spied Olympus Mons, the largest known volcano in the Solar System, and the colossal canyon system Valles Marineris, which was later named after the spacecraft.

Overall, NASA's Mariner programme was a big success. Of the 10 spacecraft launched, seven were successful! (Given Mars's track record in particular, this is a high success rate.) The Mariners were designed to perform fly-bys, probe atmospheres, explore magnetic fields and map the surfaces of

Mercury (*Mariner 10*), Venus (*1*, *2*, *5*) and Mars (*3*, *4*, *6*, *7*, *8*, *9*). The four Mariners that reached Mars (*4*, *6*, *7*, *9*) returned a wealth of information, forming our early understanding of the Red Planet.

The Martian yellow storm of 1971

By the end of 1962, the Soviet Union had attempted five launches. Given the fact that not a single one went as planned or managed to send back any data about Mars, Soviet scientists remained impressively resolute in their desire to get to the Red Planet.

Sadly, their luck didn't improve. After the ill-fated *Sputnik 24* came *Zond 2* in 1964 – a planned fly-by that ended in loss of communications – *Mars 1969A* and *1969B* in 1969 – both of which exploded shortly after launch – and *Kosmos 419* in 1971, a planned Mars orbiter that suffered a thruster malfunction causing it to fall to Earth after just a couple of days in Earth orbit*.

Finally, in 1971 (the same year as the US's *Mariner 8* and *9* launches), Soviet scientists caught a break with a duo of spacecraft named *Mars 2* and *Mars 3*. Both spacecraft were composed of an orbiter and a lander; each lander also carried a small rover that would wheel its way out on to the Martian surface after landing. Although previous US launches had given us a good idea of what Mars was like from above, no mission had yet made it down on to the planetary surface and the USSR desperately wanted to nab the title.

On 19 May 1971, *Mars 2* streaked up into the sky from Baikonur Cosmodrome in Kazakhstan, followed nine days later by *Mars 3*. Both aimed to image Mars's surface and atmosphere, and to study the planet's physical characteristics.

Mars 2 reached Mars in November of 1971 and settled into orbit around the planet. Sadly, it arrived just too late to be a

---

* A mission named *Zond 3* also launched in 1965. It's thought to be a craft originally planned for Mars that missed its launch window, causing it to be repurposed as a lunar fly-by mission.

trailblazer given the successful arrival of the US's *Mariner 9* mission a mere fortnight earlier. *Mars 3* arrived shortly afterwards.

Both arrived to an unexpected sight – a global dust storm blocking the entire surface of Mars from view! The storm was the largest one anyone had ever seen, masking all Martian features barring the very tip of the 25km-tall (16 miles) Olympus Mons and cloaking the entire planet in a thick smog. The storm had been present when *Mariner 9* arrived and showed no signs of abating.

This was a huge problem for both *Mars 2* and *3*, which were hoping to release their landers immediately. NASA had decided to delay its mission plan for *Mariner 9* until the storm ended, so that the photos sent back didn't simply show various shots of dust, dust and more dust. Unfortunately, despite the conditions, *Mars 2* and *3* didn't have the same option available to them because their computers could not be reprogrammed. Both went ahead and deployed their landers as planned, ignoring the huge storm raging below.

Mars 2 went first. Its lander separated from the orbiter and entered Mars's atmosphere, but its angle of entry was too steep and its speed too high. It struggled to deploy its parachute. It plummeted down to the surface and crashed, becoming the first artificial object to crash-land (known more sympathetically as a 'hard landing') on Mars – a dubious title.

Next up was *Mars 3*. The lander managed to release its parachute, which it used together with a mixture of aerobraking (using the drag of a planetary atmosphere to slow down) and rockets to complete a successful ('soft') landing. The *Mars 3* lander quickly flickered on, communicating with its orbiter and transmitting a view of the Martian surface. For 15–20 seconds, all seemed to work … but then, all went silent. It's thought that the dust storm knocked the small lander over, perhaps just after it managed to release its little rover, or that coronal discharge from the storm may have shorted the lander's electronics. The received image was, sadly, unusable, showing just a featureless grey background (likely because the lander hadn't yet fully deployed its imaging equipment). *Mars 3*'s lander technically became the first ever man-made object to land on another planet … even if it did only survive for a matter of seconds.

Even though these two missions weren't exactly successes, they weren't failures either. Both managed to launch, reach Mars, enter orbit, deploy their landers, and take photographs and measurements of the planet. Unfortunately, the dust storm stopped the Soviet Mars missions from returning detailed images of the planet, but the data obtained by the orbiters still represented a full cycle of scientific observations. As the orbiters could hang around for a little longer waiting for the storm to dissipate, they managed to put together relief maps of the planetary surface, explore the composition of Mars's atmosphere and more.

The colossal dust storm – dubbed the 'Martian yellow storm of 1971' due to the often ochre-like colour of Mars's dust – finally abated in January of 1972. It was in full force in mid-November of 1971, when *Mariner 9* first arrived at the planet, and was thought to have started a couple of months earlier, meaning that it raged, planet-wide, for months. Mars's surface was partially visible in mid-September and became partially visible again in mid-December, with the storm building to a peak and obscuring the entire planet in the intervening months.

While the *Mars 3* lander had experienced Martian soil between its robotic toes, there was still no successful robot operating on the surface of the Red Planet. The US threw its efforts behind a programme named Viking, a duo of orbiter–lander spacecraft that launched in 1975, based closely on the technology and design tested by the Mariner probes. The Vikings were the US's first ever attempt at landing on Mars and were the most expensive missions that NASA had ever developed.

Vikings 1 and *2* both worked perfectly! The orbiters entered orbit smoothly and the landers landed without a hitch. On 20 July 1976, the *Viking 1* lander beamed back the first clear image of the Martian surface, showing a scattering of rocky pebbles around its 'feet' – the first of many Viking photographs of the planet's surface. The landers explored the soil and rocks around their landing site, studying their composition (detecting various elements but no signs of organic molecules or life). Crucially, the photographs from the orbiters showed widespread and significant signs of potential past water activity on the planet.

The Soviet Union continued with its series of Mars probes, developing *Mars 4*, *5*, *6* and 7 in quick succession. Rather than combine their landers and orbiters into two launches, as with Viking, they instead separated them into four individual missions; *Mars 4* and *5* were planned orbiters, while *Mars 6* and 7 were landers. This design was likely motivated by the less-than-ideal positioning of the planets during the launch window, which meant that launches could only carry more limited payloads.

All four missions blasted off between 21 July and 9 August in 1973. All launched successfully and set off on their merry way to Mars.

However, arrival was not to be as easy. The missions had been developed relatively hastily in a rush to beat the two planned Viking launches, and given the green light despite a known problem with their on-board electronics. Just months before launch, ground tests detected a problem with some of the microchips used in the spacecraft's construction, noting that they deteriorated and corroded over time and were likely to fail after 1.5 to 2 years of operation – a timeline that coincided with the missions' expected arrival at Mars. Overall, the chips' odds of succeeding were estimated at just 50 per cent.

The decision to launch proved costly. A faulty microchip failed to trigger *Mars 4*'s rockets at arrival, meaning that the orbiter couldn't get close enough to enter Martian orbit. Ground control managed to switch on the probe's cameras for a few minutes during its closest approach, obtaining some imaging data and a couple of surface panoramas. After its short-lived period of operations *Mars 4* flew serenely on, entering orbit around the Sun.

The second orbiter, *Mars 5*, didn't have the same problem and managed to settle into orbit around Mars as planned.

However, a problem quickly arose: the probe's main instrument compartment slowly began to depressurise, limiting its possible lifespan to just two or three weeks. Scientists quickly designed an accelerated imaging programme for the spacecraft and *Mars 5* got to work. Despite its problems, *Mars 5* provided us with some useful data: it sent back 43 usable photographs that showed clear signs of water erosion, adding to the case for past water on Mars, and an additional five surface panoramas. On 28 February 1974, after a handful of key imaging sessions, *Mars 5* went silent.

The landers followed a strikingly similar pattern: one partial success, one complete failure. *Mars 6* began its descent to the Red Planet as planned, travelling down through the atmosphere and beaming back information about its surroundings as it did so. It then stopped speaking abruptly during its landing sequence. The data it did manage to transmit were completely garbled and fittingly alien … due to a faulty microchip.

Mars 7 didn't even manage to get that far. The lander separated from its transport craft too early, missing Mars by 1,300km (800 miles). The reason? You guessed it: a faulty microchip.

While many of these launches showed glimmers of success, the only pre-1976 missions generally classed as wholly successful are those belonging to NASA's Mariner and Viking programmes. The following years saw a multitude of launches that were failed, successful, or somewhere in between.

The final launches attributable to the Soviet Union (dissolved in 1991 to become the Russian Federation) were in July 1988 in the form of *Phobos* (or *Fobos*) *1* and *2*, both of which went radio silent. These missions were named in honour of Mars's largest moon, and carried small landers that would have hopped down on to its surface. In 1996 Russia launched *Mars 96*, which sadly didn't get very far – it entered orbit around the Earth before suffering thruster problems and crashing down somewhere near Bolivia. After *Mars 96* Russia dialled down its efforts to get to Mars. Its next launch was some 15 years later, leaving plenty of room for others to take the lead.

The last of the Viking orbiters and landers officially retired in November of 1982, leaving the US space programme with a prominent Mars-sized gap. Although the Vikings had uncovered an incredible amount about our red neighbour, there was still much to explore and much we didn't know (and still don't!). In the years between the Vikings' retirement and 2013, the US developed and launched 14 NASA-led missions – an average of one Mars mission every two and a bit years.

This burst of activity started with *Mars Observer* in 1992, a geoscience and climatology orbiter mission that went radio silent before reaching Mars. This was a huge problem for NASA, which had spent over US$800 million developing the spacecraft. The failure of *Mars Observer* prompted NASA to step back, reassess and put together a more pragmatic long-term plan known as the 'Mars Exploration Program'. This encompassed all of the goals we still currently hold – detecting water, determining if life ever existed on the planet and, if so, what kind of life, characterising the Martian atmosphere and climate, and understanding its geology and surface features. Overall, the programme aimed to find out what Mars is like and how it came to be so – essentially, how and why it's different to Earth.

The *Mars Observer* fiasco also triggered a change of attitude. Then-NASA administrator Daniel Goldin championed a new philosophy of 'faster, better, cheaper' (FBC). This was reflected in the organisation's 'Discovery Program', founded in 1992, which comprised a set of low-cost missions dedicated to exploring the Solar System. The programme aimed to promote the use of more affordable technologies in cost-capped space launches. Targets included asteroids, comets, the Moon, the solar wind, Mercury, Venus and Mars. Later missions extended this scope beyond our local neighbourhood to include other science goals, such as the study of extrasolar planets.

The formation of the FBC mentality and the Discovery Program was at least partially based on the high risk involved in trying to get to places like Mars. It just wasn't feasible in the long term to spend a billion dollars on a spacecraft that might just end up in pieces at the bottom of an ocean (and on

Earth, no less). NASA had spent *a lot* of money trying to get to Mars and a huge amount of that had been wasted. Even the successful missions had cost a remarkable amount. For example, the Viking missions cost almost US$1 billion to develop (in 1970s money – over US$4.5 billion now). NASA described the reasoning behind their FBC approach in a 2000 report, stating that 'we have tried to do too much, too fast. We have lost sight of how difficult space exploration is and will have to slow down … we have had too many mission failures which must be fixed'.

There was one pre-FBC mission in development at the time of the *Mars Observer* failure: NASA's *Mars Global Surveyor* (*MGS*), launched in 1996. As the name suggests, *MGS* was designed to map the entire planet and its atmosphere from orbit.

MGS was hugely successful, somewhat mitigating the dismay over *Mars Observer.* The probe arrived at Mars, entered an elliptical orbit on 12 September 1997 and began slowly trimming its orbit to a circular near-polar one before starting work in March 1999. Originally *MGS* was designed to study Mars for a complete Martian year, but it surpassed all expectations. After completing its primary objectives, *MGS* gained three successive mission extensions before going radio silent in late 2006.

As well as mapping the surface of Mars, *MGS* targeted the planet's surface features, geology, surface composition and properties, global topography, shape, gravitational and magnetic fields, weather and atmospheric structure, and studied the interaction between Mars's surface and atmosphere. In its extended phases, *MGS* went on to scope out potential landing sites for future NASA missions, and continued long-term monitoring of the properties both on and around Mars. All in all, the industrious probe sent back more images and science data than all previous Mars missions combined.

Enter the rovers

So far, all of these missions have been similarly composed of orbiters and landers, some containing small rovers that never quite made it on to Martian soil. However, shortly after the launch of MGS, the US decided to change the game. In 1996, NASA sent up the first of Mars's most famous robotic residents – a rover by the name of *Sojourner*.

Sojourner hitched a ride aboard *Mars Pathfinder*, a mission comprising both the six-wheeled surface rover and a stationary lander. *Pathfinder* was the first Mars-bound NASA mission to launch as part of the faster, better, cheaper Discovery Program, with just three years of development time and a cost capped at US$150 million.

On 4 July 1997 the craft sped down through the Martian atmosphere and landed in a region named Ares Vallis ('Mars Valley'). It used a parachute to slow down during its descent and a giant set of multiple heavy-duty airbags to cushion its landing. It hit Mars's surface, bounced high into the air over and over again for a good couple of minutes, and finally came to a stop about a kilometre from its impact site. It then unfurled to release *Sojourner*, the first robot to ever set wheel on another planet.

Originally known as the *Microrover Flight Experiment* (*MFEX*), *Sojourner* was named after Sojourner Truth, an African-American women's rights and anti-slavery activist who lived in America in the early 1800s. Born Isabella Baumfree, Truth adopted her moniker – 'Sojourner' meaning 'traveller' – after escaping slavery and deciding to 'travel up and down the land' preaching about abolitionism and promoting equal rights for all citizens. The inspiration and name were proposed by 12-year-old Valerie Ambroise from Connecticut, US, via an essay competition launched by NASA in 1995.

Sojourner weighed in at 11.5kg (25lbs), had six wheels and was topped by a flat panel covered in solar cells. She was roughly the size of a milk crate or microwave oven. The rover was designed to carry out a primary mission lasting seven '*sols*' – each *sol* being a full Martian solar day lasting 24 hours, 39 minutes and 35 seconds – with a possible extension to 30 *sols*, all being well.

Sojourner worked even better than planned and survived far past the 30-*sol* extension. For 83 *sols* (85 Earth days) the rover cautiously explored Mars, sticking close to the *Pathfinder* lander. During its tenure on Mars the rover returned valuable information about Mars's soil, rocks, geology, magnetism and atmosphere, and snapped hundreds of photographs of its surroundings. When it finally retired, *Sojourner* had travelled an overall distance of just 100m (330ft)!

Sojourner also analysed the Martian soil at 16 different locations and explored the rocks sitting around its landing site. Scientists named these rocks after various famous cartoon characters based on their appearances; 'Barnacle Bill' had a rough surface with barnacle-like features running across it, 'Yogi' resembled the back of a bear's head and so on. *Sojourner*'s mission geologists evidently had fun naming all of the rocky features shown in the photographs sent back by *Pathfinder;* all had light-hearted names, including Marvin the Martian, Calvin, Hobbes, Indiana Jones, Garfield, Space Ghost, Jedi, Shaggy, Scooby Doo, Flipper, Rolling Stone, Poptart, Warthog, Dilbert, Dilbert's Boss, Elvis, Poohbear, Piglet, Asterix, Obelisk, Snoopy, Tigger and Darth Vader.

In 1997, the *Mars Pathfinder* lander was renamed as the '*Sagan Memorial Station*' in honour of late astronomer Carl Sagan, who passed away in 1996. 'Carl Sagan was a very unique individual who helped young and old alike to dream about the future and the possibilities it may hold,' said then-NASA administrator Daniel Goldin in a press release. 'Carl always liked to push the boundaries too, and the *Mars Pathfinder* mission, with its rover *Sojourner*, clearly has done that.'

After the roaring success of *Mars Pathfinder* and *Sojourner,* Japan decided it was time it joined the Martian Space Race.

In 1998, the Japan Aerospace eXploration Agency (JAXA) launched *Nozomi*, an orbiter previously known as 'Planet-B' (*Nozomi* is Japanese for 'hope' or 'wish'). The spacecraft intended to study Mars's upper atmosphere, a field of science known as aeronomy, and analyse how it interacted with the solar wind and magnetic field. *Nozomi* launched successfully, swung around the Moon and then back around the Earth and burned its thrusters, attempting to settle on to a Mars-bound trajectory. However, while swinging around the Earth one of the valves in the spacecraft malfunctioned. *Nozomi* subsequently lost fuel and was unable to reach its desired flight path.

JAXA quickly reworked its mission plan. Rather than reaching Mars in October 1999 as planned, *Nozomi* instead entered an orbit around the Sun. The plan was for the spacecraft to perform two gravitational manoeuvres around Earth to slow it down slightly, one in 2002 and one in 2003, so that it could try to encounter Mars in December 2003. However, during the first Earth fly-by in 2002, a solar flare damaged *Nozomi*'s on-board power systems and electronics, causing the heating system to malfunction and the fuel to freeze solid! Luckily, as *Nozomi* neared Earth, its fuel thawed enough that the probe could successfully slingshot around our planet to correct its trajectory as planned. The second fly-by was also successful, as *Nozomi*'s proximity to the Sun in the intervening months had completely thawed its fuel reservoir, and the spacecraft set off for Mars several years later than intended. Sadly, the unfortunate mission again encountered an unexpected failure when scientists were unable to correct its orientation, meaning that it couldn't fire its thrusters correctly to enter Mars orbit. After countless attempts to salvage it, the poor and unlucky *Nozomi* was abandoned.

NASA also struggled to recreate the success of *Sojourner*. The organisation launched the *Mars Climate Orbiter* (*MCO*) in 1998, followed by the *Mars Polar Lander* (*MPL*) in early 1999. The former came far too close to Mars while attempting to orbit the planet, skimming just 57km (35 miles) above the surface – roughly 23km (14 miles) closer than the spacecraft could endure, and a whopping 170km (105 miles) lower than planned. As a result, *MCO* passed through the planet's upper atmosphere, burning up and disintegrating under the intense atmospheric stress.

MCO failed due to an incredibly frustrating case of human error. NASA determined that 'the root cause for the loss of the *MCO* spacecraft was the failure to use metric units in the coding of a ground software file used in trajectory models' – in other words, part of the spacecraft's code was speaking in feet while the rest was speaking in metres, causing *MCO* to shift closer to Mars with each orbital manoeuvre until its path was unsalvageable.

NASA's second attempt, the *Mars Polar Lander* (*MPL*), didn't suffer from such a toe-curling mishap, but didn't fare much better than its forebear. Launched on 3 January 1999, *MPL* aimed to land on a region of the Martian surface near Mars's south pole and explore its climate and geology. It also carried two piggy-backing grapefruit-sized probes, named *Scott* and *Amundsen* after two of the first explorers to arrive at the South Pole, that would detach and hurtle down to hit Mars's surface and penetrate deep beneath its surface (aptly known as 'penetrators' or 'impactors'). *MPL* and its cargo entered Mars's atmosphere … and disappeared without a trace.

Fortunately, the aforementioned *Mars Global Surveyor* (*MGS*), launched in 1996, was still operating in Mars orbit. Scientists used *MGS* to scour the Martian surface for signs of *MPL*, but found no sign of it. A mission post-mortem determined that *MPL* deployed its parachute and fired its landing rockets as planned, only for the spacecraft's on-board systems to mistake the jolt caused by the deployment of the landing legs for ground contact. The rockets then switched off, despite there still being 40m (130ft) between *MPL* and the surface, causing the poor little lander to tumble to the ground and break apart.

In 2005, planetary scientist Michael Malin (of Malin Space Science Systems, supplier of *MGS*'s onboard camera), revisited the *MGS* photographs from 1999 and early 2000. Using more recent knowledge and experience in monitoring spacecraft on Mars, Malin claimed to find features in the planned landing zone that resembled a parachute, potential impact site and disturbed ground. 'It seems that the *MPL* investigation board may have been correct,' wrote Malin in *Sky & Telescope* in May 2005. '*MPL*'s descent proceeded more or less successfully through atmospheric entry and parachute jettison.

It was only a few short moments before touchdown that disaster struck.'

None of the missions mentioned thus far are still operational but the next chronological launch after *MPL*, in 2001, is still happily working away in orbit around Mars, bringing us from the past into the present.

This success was NASA's *2001 Mars Odyssey* mission, named as a tribute to the famous science-fiction novel *2001: A Space Odyssey* by Arthur C. Clarke. The mission was originally named *ARES*, short for *Astrobiological Reconnaissance and Elemental Surveyor*, but the name was changed after officials – including NASA's Mars Program Director Scott Hubbard – felt the name was 'not very compelling, awfully aggressive, and as an acronym pretty nerdy'. The name '*2001 Mars Odyssey*' had been initially rejected due to concerns over copyright, but NASA decided to reach out to Clarke himself to see what he thought. He 'enthusiastically endorsed the name', replying with 'no objections … in fact, [he was] delighted by the idea'!

2001 Mars Odyssey aimed to hunt for evidence of water on Mars, as well as continuing the tasks of its ancestors in exploring the planet's surface and characteristics (atmosphere, radiation, composition). The probe has helped to characterise the composition of the Red Planet's soil and surface, detected water lurking beneath Mars's surface and calculated estimates for Mars's radiation environment. It is still operational, winning it the title of the longest continually operational spacecraft orbiting a planet other than Earth (quite a fitting achievement given its moniker). The probe has also helped to scope out potential future landing sites for NASA from its unique vantage point and has acted as a communications link for other missions on Mars's surface.

Following *Odyssey*, NASA returned to the Red Planet with two new rovers, named *Spirit* and *Opportunity*, brought together under the moniker of the Mars Exploration Rover Mission (MER). MERs A (*Spirit*) and B (*Opportunity*) launched on 10 June and 7 July 2003 respectively, and landed successfully in two different regions of Mars in January 2004.

These rovers bore little family resemblance to *Sojourner.* At 185kg (408lbs) each, *Spirit* and *Opportunity* were over 15 times more massive than their predecessor and far larger, with a height of 1.5m (4.9ft), width of 2.3m (7.5ft) and length of 1.6m (5.2ft). Both had six wheels and two folded 'wings' covered in solar panels. The MER rovers were the first to take the characteristic 'rover' appearance continued by *Curiosity* some years later: wheels supporting a boxy body, extendable 'arms' for scientific experiments and a 'head' comprising a set of panoramic cameras supported on a thin 'neck'. For *Spirit* and *Opportunity*, both of which were about the size of a golf cart, their 'necks' elevated their cameras to nearly eye level.

The rovers were sent to two different sites on the Martian surface, both in the planet's equatorial regions. *Spirit* landed in Gusev Crater and *Opportunity* touched down at Meridiani Planum, both sites thought to have potentially once hosted liquid water. Both rovers got to work studying their surroundings; they explored the geology of their Martian habitat and studied individual rocks, soil samples and surface features. From the data that *Opportunity* has sent back, we believe that its landing site was once completely flooded with water and, according to NASA in 2003, 'was once the shoreline of a salty sea'.

The robotic duo also returned a great many photographs of the Red Planet. Both outlasted their proposed mission timeline by far – in fact, *Opportunity* is still active today. The rover is still making its way across the rusty Martian surface, studying the planet's geology and environment, analysing rocks, minerals and soil, and searching for evidence of past water and microbial life. Its deceased sibling remains quietly sitting on the surface of Mars, parked in the spot where it last spoke to Earth.

As NASA was preparing to launch its duo of rovers, Europe was preparing to join the race for the first time. In 2003, the European Space Agency (ESA) launched its *Mars Express* mission, comprising the eponymous orbiter and a lander named *Beagle 2* (funded and built by the UK). *Mars Express* was hugely successful and is still in operation today. It was built quickly – it was developed in just five years, compared with the roughly 10-year lead-up needed for previous similar missions – and managed to take advantage of a launch window where Mars was especially close to Earth, meaning that its speedy journey took just six months. These factors fed into the primary mission's name, '*Express*', while the lander took its moniker from the name of Charles Darwin's famous ship, HMS *Beagle*.

Mars Express primarily aimed to search for signs of subsurface water on Mars, which could render the planet friendly to life. It also began to map the Martian surface and atmosphere, figuring out what both were composed of. It explored Mars's polar caps, finding them to contain possible water ice, and detected traces of methane and ammonia in the atmosphere. These compounds are particularly interesting, as described in the previous chapter, because they are common signs of either recent geological or biological activity.

The mission's lander, *Beagle 2*, had worse fortune. It set off happily, successfully detaching from *Mars Express* on 19 December 2003. It was scheduled to touch down on Mars's surface on Christmas Day, but it never phoned home. Everything seemed to have gone as planned, but *Beagle 2* just disappeared without a trace. Its whereabouts remained completely unknown for over a decade until the little lander was eventually found in 2015 images from NASA's *Mars Reconnaissance Orbiter*, launched a couple of years after *Beagle 2*. The probe can be seen, apparently intact, sitting in the middle of Isidis Planitia with its parachute, landing gear and outer coverings discarded nearby! *Beagle 2* was designed with solar panels arranged as stacked petals, which would slowly flip out one by one to reveal the main body of the lander. It's thought that one of *Beagle 2*'s solar panels failed to fold open, obstructing the lander's instruments and, crucially, its communications system, meaning that it was unable to speak to Earth.

The next decade saw a flurry of Mars-related activity. ESA again flirted with Mars some years later in 2007 using an unlikely probe: *Rosetta*. Launched in 2004, *Rosetta* was en route to study, and land on, a comet named 67P/Churyumov–Gerasimenko, something it successfully managed to do in 2014. While it was winging its way to its cometary target, *Rosetta* performed a fly-by of the Red Planet, returning a beautiful view of Mars snapped from just 1,000km (620 miles) up. NASA's *Dawn* spacecraft also performed a successful fly-by of the planet in the summer of 2009, while on its way to explore the asteroid belt.

NASA continued its streak of good fortune and technological success with its *Mars Reconnaissance Orbiter* (*MRO*, 2005) and *Phoenix* (2007) missions. *MRO* is still functioning in orbit around Mars, while *Phoenix* landed in Mars's northern polar regions and studied them for over 150 *sols*. Together, *MRO* and *Phoenix* discovered an incredible amount about Mars, especially concerning water. They measured the amount of water ice in Mars's polar caps, found ice deposits, chlorides and phyllosilicates (both mineral types thought to have formed in the presence of water) in various craters and surface features, photographed a variety of geological processes, and explored the planet's weather and climate cycles.

Unfortunately Russia fell back into its old pattern with its first attempt in a long while, 2011's *Phobos-Grunt* (or *Fobos-Grunt*, Phobos being Mars's largest moon and 'grunt' being Russian for 'soil'). China's first probe aimed at Mars, *Yinghuo-1* (meaning 'firefly'), was launched alongside *Fobos-Grunt* – both sadly fell back to Earth due to a programming error with the propulsion system, tearing down through our atmosphere and falling apart.

Why so curious?

While these missions have completely revolutionised our view of the Red Planet over the past half a century, one mission has captured public imagination like no other.

On 26 November 2011, NASA launched its most ambitious mission so far: *Mars Science Laboratory* (*MSL*). Shaped a little like a spinning top, *MSL* weighed in at nearly

4 tonnes (8,800lbs)! The spacecraft carried a set of instruments for monitoring deep space while en route to Mars (which were later used as guidelines for potential human travel to Mars, given the dangers of deep-space radiation), and an outer shell and gigantic heat shield locked together like a huge metallic Kinder Egg. *MSL* was carrying precious cargo inside this little capsule. Wrapped up with its own landing equipment was the most advanced robotic rover yet developed: *Curiosity*.

Curiosity alone weighed 900kg (1,985lbs). With no familiar landmarks or scales on Mars to use for reference, the rover's size may come as a surprise. The word 'rover' brings to mind something relatively small, but *Curiosity* is anything but. *Sojourner* was, as mentioned, the size of a large remote-controlled car, while *Spirit* and *Opportunity* grew to the size of golf carts. *Curiosity* carries a bulk and height more akin to that of a small car, towering over its human creators at 2.2m (7.2ft) tall. Somehow, NASA had to land this giant rover on the surface of an alien planet notorious for eating spacecraft for breakfast.

Curiosity aimed to land in Gale crater, a landing site that has since been named 'Bradbury Landing' after the Mars-loving sci-fi author Ray Bradbury. The rover's landing sequence, which took place on 6 August 2012 and was automated due to the Earth–Mars time difference, was tellingly dubbed 'seven minutes of terror'. This was especially stressful due to the Earth–Mars time delay; by the time we were alerted to *Curiosity* entering the Martian atmosphere the rover would already be down on the planet's surface – alive or dead.

MSL approached Mars, reaching just 125km (78 miles) from the planet's surface. Then, its landing sequence began. It separated from its cruising equipment and descended through Mars's atmosphere, plummeting down, down, with the heat shield bearing the brunt of the fall and enduring scorching temperatures of over 2,100°C (3,800°F). The atmosphere slowed *MSL* as it fell, reducing its speed from its initial deep-space velocity of over 21,200kmph (13,200mph) to some 1,600kmph (1,000mph) – far slower, but still nowhere near slow enough. When *MSL* reached 11km (7 miles) from the ground its parachute deployed, instantly decreasing the spacecraft's speed to just 9 per cent of its original velocity. A few kilometres later *MSL*'s heat shield detached, popping off and giving *Curiosity* its first ever glimpse of Mars. At just 1.6km (1 mile) up the rover detached from its parachute to fend for itself, free-falling through the air at 290kmph (180mph).

Curiosity was packed alongside its landing gear, which comprised a set of retrorockets and the futuristically named Sky Crane. The retrorockets fired during its final kilometre of free-fall, slowing the rover down further – and then the Sky Crane jumped into action. It essentially acted as a docking station, a crane, from which it lowered the rover to the ground, using its retrorockets for stability. *Curiosity* dangled precariously from the crane, held by a set of nylon ropes. As with an aeroplane, the rover began to quickly lower its wheels and arrange its landing gear, just in time for the Crane to detach and fly away in order to crash some distance away from the rover. This may sound like the plot of a science-fiction film, hence scientists' anxiety about the operation, but it all went smoothly – unbelievably smoothly, in fact, with the rover landing under 2.5km (1.5 miles) from its intended bullseye.

The mission's operations have been just as successful as its landing. So far the rover has roughly doubled its intended lifespan and is still going strong. *Curiosity* has poked and prodded its way around Gale crater and its central mound, Mount Sharp (which appears to have well-preserved rock layers from ancient times to present day) looking for signs of water and past life, and studied the Red Planet's weather and climate systems. More generally, *Curiosity* is hunting for possible habitats for microbial life and scoping the area for possible locations for a human base (more on this later). Arguably the rover's key finding has been that its landing site is, or was, habitable – a huge discovery.

'As the mission's gone along, it's only gotten better,' says Ashwin Vasavada, the current NASA project scientist for

MSL/Curiosity. 'We've found more and more habitable environments as we've been climbing Mount Sharp. Gale Crater was a friendly place for life three billion years ago – that's the bottom line.'

As an interesting aside, the mission has also been a huge success in terms of public engagement. The rover has been thoroughly anthropomorphised and even has a Twitter account, ostensibly manned by the quirky rover itself. This account recently revealed that, heartbreakingly, *Curiosity* quietly sings itself 'Happy Birthday' each year! The public has responded incredibly well to *Curiosity*'s public persona. '*Curiosity* had the benefit of learning a lot from a previous Twitter account for @MarsPhoenix (2008),' explained Veronica McGregor of the *Curiosity* social media team in a 2013 Reddit 'Ask Me Anything'. 'It was obvious during @MarsPhoenix that using first person was the best way to go. People were more responsive to the first [person] and it was easier to fit tweets into 140 characters. *Curiosity* is a mash-up of personalities from three of us working together each day. We want to make it fun but educational and interactive.'

The robots of the future

Prior to 2013, only three space agencies had managed to get to Mars – the Soviet space programme, NASA and the European Space Agency. This all changed with the launch of the Indian Space Research Organisation's (ISRO) *Mars Orbiter Mission* (*MOM* but also named *Mangalyaan*, 'Mars-craft'). This probe was the first Mars-bound mission developed by India and, despite it being the organisation's first ever launch to the Red Planet, all went smoothly. The probe launched on 5 November 2013 and settled into orbit around Mars on 24 September the following year. The spacecraft primarily aimed to demonstrate and develop many of India's launch and spaceflight technologies, with a secondary focus on exploring the characteristics of Mars's atmosphere and surface.

MOM is notable for another aspect – its cost. The mission was remarkably cheap. Its reported budget clocked in at just

4.5 billion rupees, the equivalent of US$74 million at the time. For comparison, *The Martian* had a budget of US$108 million. India managed to actually get to Mars using less money than it took 20th Century Fox to make a film about it! Former ISRO chairman Kopillil Radhakrishnan attributed this cost-effectiveness to a mix of frugality, low labour costs, modular technology, a push towards low-cost innovation and tight schedules.

In 2013 NASA launched its *MAVEN* (Mars Atmosphere and Volatile EvolutioN) mission, which entered orbit around Mars on 22 September 2014 (and cost around ten times more than ISRO's *MOM* – an oft-made comparison given the probes' similar launch dates). *MAVEN* had a specific mission: to understand Mars's atmosphere! What is it made of? How does it interact with surrounding space? Where did it all go? As NASA describes, *MAVEN* data aimed 'to determine the role that loss of volatiles from the Mars atmosphere to space has played through time, giving insight into the history of Mars's atmosphere and climate, liquid water, and planetary habitability'. *MAVEN* was responsible for the aforementioned discovery about Mars's leaking atmosphere (that the solar wind is slowly stripping it away from above).

Last year saw the launch of the first of two ExoMars missions, a collaboration between the European Space Agency and Russia's Roscosmos (part of the European-led ExoMars project). This mission has two phases – firstly an orbiter and test landing module (*ExoMars 2016*, launched in March 2016), and secondly a rover and surface platform (*ExoMars 2020* which, as the name suggests, will launch in 2020).

ExoMars 2016 is the heaviest spacecraft we've sent to Mars. It comprised the *Schiaparelli* Entry, Descent and Landing Demonstrator module (EDM), which aimed to head down to Mars's surface to test its equipment, establish an initial communications base and study its surroundings for a few *sols*, and the *ExoMars Trace Gas Orbiter* (*TGO*), which will study the Martian atmosphere from orbit. At the time of writing *ExoMars 2016* had just arrived at Mars (October 2016). The *Schiaparelli* lander separated from *TGO* as planned and entered the Martian atmosphere. It gathered data during the first part of its descent before going unexpectedly silent – sadly, it is believed to have crash-landed on Mars. *TGO*, however, successfully entered orbit around Mars, and all looks good for future operations – including, crucially, its support of the upcoming *ExoMars 2020* mission when it arrives at the Red Planet.

This brings us to the present day! Upcoming missions, both public and private, will be the robots and Mars explorers of the future (more on this later). At the time of writing there are eight robots operating at Mars, either on the surface or in orbit. If current mission plans are to be believed, these numbers are to steeply increase in the next decade or two.

Mars had better prepare itself for a never-before-seen spike in interplanetary tourism – and a whole host more visitors.

CHAPTER ELEVEN

Destination: Mars

Space travel is hard, but it seems that getting to Mars is far harder than it should be.

Of all the spacecraft we've sent to Mars, only a third or so have been fully successful. This failure rate has caused NASA officials to jokingly refer to a 'Martian curse', and journalists to dream up the idea of a 'Great Galactic Ghoul' – a ravenous space monster that hungrily gobbles up the probes we send its way, periodically awakening to fiercely defend its territory from alien invaders.

However, we've sent numerous missions to both the Moon and the *International Space Station* (*ISS*), both manned and unmanned, and have been doing so for so long that it's almost a matter of routine in the latter case. Why is getting to Mars so challenging?

Distance

The first issue is simple: Mars is much further away from us. One of the key problems in travelling beyond our little patch of space is the sheer amount of travel time involved. The Solar System may be small on a cosmic scale, but it's immense on a human one.

The distance from the Sun to the outermost planet, Neptune, is around 4.5 billion km (2.8 billion miles), and the only mission to have visited Neptune, *Voyager 2*, took 12 years to get there. When we travelled to the Moon, we 'only' needed to traverse 384,400 km (238,800 miles). On average, Mars is located around 225 million km (140 million miles) away from us, but when it's on the opposite side of the Sun it can be over 400 million km (250 million miles) away. It may be our neighbour, but it's by no means near.

Using current technology, an average one-way trip to Mars takes around eight months, or roughly 240 days. However, as with the planet's distance from us, this time varies; past missions to Mars have taken anywhere from a speedy 128 days (*Mariner* 7 in 1969) to a more sluggish 333 days (*Viking 2* in 1975). The best and fastest time to launch is when Earth and Mars are optimally aligned in space – at 'opposition'. As mentioned previously, this alignment happens roughly every couple of years for Mars. This is where Earth sits directly between the Sun and Mars, with the trio arranged in a line Sun–Earth–Mars, and the Earth–Mars distance is lower than at any other point in orbit.

Because planetary orbits aren't perfect circles, some oppositions bring Mars closer to Earth than others. The optimum time is therefore when Mars is at opposition but also at its closest orbital approach to Earth – this aligns with Mars's closest approach to the Sun, an orbital point known as 'perihelion'. This killer combination (a 'perihelic opposition') happens every 15–18 years, so is far less common. The next opposition of this kind is in 2018 and the next after that will occur in the 2030s (hence why these two time periods are being targeted by various Mars mission plans). For launch windows during opposition, we need to use less energy – sometimes half that needed at other times – to launch and travel to Mars, so missions are generally lighter, quicker and cheaper.

To get to the planet quickest, it stands to reason that one could launch a spacecraft, calculate where Mars would be in space an appropriate length of time later and then burn the craft's rockets continuously towards that location. This would indeed be by far the quickest way to get to the planet, but the amount of fuel needed would be prohibitively heavy and expensive.

Instead, we usually carry out our missions by using carefully planned orbital manoeuvres and trajectories such as a 'Hohmann transfer orbit'. Rather than heading directly for Mars, the spacecraft is launched into an orbit around the Sun that is wider than Earth's, such that it'll eventually intersect with Mars's orbit. All the spacecraft needs to do is perform an initial fuel burn to give it a burst of acceleration and then coast the rest of the way, using very little fuel in the process. Once the craft reaches Mars it uses its remaining fuel to burn rockets in the opposite direction to its motion (retrograde rockets, or retrorockets), slowing itself down relative to the planet so that Mars's gravity can essentially capture it.

While this method puts the amount of fuel needed at a minimum, it certainly isn't the quickest way to get to Mars, and travel times are between eight and nine months on average. This is also from a one-way robotic perspective; when you pack a spacecraft full of humans, you also need to consider how to bring them back. A return mission would have to grab a similar transfer window back, meaning that it'd need to hang around at Mars for months before beginning its journey home.

Launch and propulsion

We obviously already have the ability to get to Mars successfully, but future trips to Mars (manned ones in particular) call for more powerful and innovative methods of propulsion. In 2015, NASA administrator Charles Bolden told reporters that he'd like to reduce the travel time to Mars as a priority, encouraging the development of 'game-changing' propulsion methods. 'Right now it's about an eight-month mission,' he said. 'We'd like to cut that in half.'

To dramatically decrease travel time to Mars, we'll need new methods of propulsion to allow us to both carry less fuel and move faster – perhaps nuclear (using the energy released by nuclear reactions to heat propellant), electro-magnetic ion and plasma drives (which use electric and magnetic fields to drive acceleration), or solar-electric propulsion (gathering energy via solar panels, converting to electricity, and heating and ionising gas before shooting it outwards to create thrust – this is already well-developed, but we need to increase its strength for the method to be viable). Other as-yet-undeveloped ideas include antimatter rockets (which could be the fastest ever created, combining electrons and positrons to reach Mars in weeks) and laser propulsion (which could use 'photonic propulsion', the momentum of light particles, to get to Mars in mere days).

While some of these are theoretical, others are far from it. For example, nuclear rockets have been in the pipeline for NASA since the 1950s – NASA planned to send a nuclear-powered mission to Mars in the late 1980s before the plan was scrapped – and both NASA and Roscosmos have used increasingly advanced forms of ion propulsion for years. However, some are more widely applicable than others. It's possible that we could use a mixture of advanced and current propulsion methods to transport different kinds of cargo; after all, only humans would need to get to Mars speedily (more on this later). Ion propulsion, for example, is only suited to in-space propulsion, so we'd need to use some other kind to get our spacecraft off Earth's surface.

Some have suggested storing a cache of cryogenic propellant in orbit (a 'propellant depot') around the Earth, thus lowering the amount of fuel (and mass) a mission would need to carry with it during launch. Others have proposed that we build a Mars-bound craft in space, sending multiple smaller pieces into orbit via successive launches so that it can all be pieced together in situ (this is primarily relevant when considering manned missions, which will be heavier and larger in scope than any robotic mission we've sent to Mars). We used this process for the *ISS*, sending pieces up over the course of two decades and over 30 flights. The *ISS* was simply too massive to launch from the Earth, regardless of propulsion method. This approach could be well-suited for a mission architecture involving a 'mothership' that stays in orbit around Mars while smaller missions shuttle down to the surface from within, a bit like a massively scaled-up version of the Apollo missions (but it would take a long time to build and many launches).

Multiple space agencies (both governmental and private) have existing launch technologies that they're planning to scale up and advance in coming years to tackle this issue, including NASA's huge *Space Launch System* rocket and the *Falcon Heavy*, a heavy-lift rocket under development by private organisation SpaceX (more on this later). Although it's a crucial component of the trip, transportation and propulsion is unlikely to be a serious limitation in our journey to Mars (although it is the *key* limitation if we dream of reaching further out into the Universe). Even if we don't have the means just yet, it's a promising work in progress.

Landing

Landing at Mars is tricky. Mars's atmosphere is far thinner than Earth's, meaning that it's difficult to slow down. Mars has just enough atmosphere that it needs to be accounted for in our mission plans, or it'll heat up and tear apart any incoming spacecraft, but it doesn't have quite enough for us to rely on it as the sole method of braking or controlling the speed of descent ('atmospheric braking' or 'aerobraking'). To add another layer of difficulty, Mars is an unfriendly place to land. It's dusty and unpredictable and rough, with jagged craters and slopes and rocks and lots of sharp edges that could easily puncture an airbag.

This means that we have to develop more complex methods of navigating the Martian skies and touching down safely (something known as EDL – 'entry, descent and landing'). Scientists are developing lighter, bigger and more effective heat shields – such as NASA's giant, conical Hypersonic Inflatable Aerodynamic Decelerator (HIAD) – which could help to slow spacecraft down significantly.

There are a few main ways to land on Mars, all of which require some kind of heat shield and parachute but have slightly different landing sequences. Once it begins, we have no control over the EDL sequence at all: it's completely automated and controlled by the spacecraft's on-board computer systems. Due to the time delay between Earth and Mars, even when we get the notification that the probe has entered Mars's atmosphere, it may already have crashed or suffered some other awful fate on the surface below.

The *Spirit* and *Opportunity* rovers, which landed on Mars in 2004, relied on atmospheric braking for the first four minutes of their descent, unfurled parachutes once they hit the altitude of a typical aeroplane flight path and then burned retrorockets. They still had to endure a terrifying period of free-fall from four storeys up, with only airbags and cushioning to soften the blow. Each hit the Martian surface at a speed of 48kmph (30mph), and bounced around 15 times before coming to a halt.

As described previously, *Curiosity* upped the ante and carried its own Sky Crane to Mars, as the rover was too heavy and intricate to rely on bouncing airbags. This allowed it to touch down more gently than either of its cousins, but made for a more nail-biting landing for the poor scientists stuck here on Earth. As our missions become heavier, more complex or manned, we will need even gentler and more reliable methods of landing.

On the Martian surface

Mars has a surface only a geologist could love. The powder-covered surface and dust-permeated air on the planet is unfamiliar to us; it may contain chemicals that humans find toxic, that irritate or even burn the skin, or could clog our lungs and airways.

Apollo astronauts reported that lunar regolith triggered irritative symptoms similar to a head cold or allergy, causing them to complain of 'lunar hay fever'. NASA says that Martian dust 'could be even worse. It's not only a mechanical irritant but also perhaps a chemical poison … the dusty soil on Mars may be such a strong oxidiser that it burns any organic compound such as plastics, rubber or human skin as viciously as undiluted lye or laundry bleach'. Luckily, there are various technologies we can use to combat this – including thin film coatings on equipment and electrostatic techniques to repel dust from clothing – but it's yet another mark on Mars's unfriendliness tally.

This caustic blanket of ultra-fine dust also means that our missions need to carefully consider their hardware and chosen power source in order to survive. Earlier rovers (*Sojourner*, *Spirit* and *Opportunity*) were sent to Mars with solar panels so they could draw their required power from the Sun. *Curiosity*, though, uses a kind of nuclear power source known as a radioisotope thermoelectric generator (RTG), which draws power from the natural decay of a radioactive isotope of plutonium (plutonium-238). This substance spews out heat as it decays, which *Curiosity* then turns into electricity to power its instruments – it's essentially a nuclear battery.

NASA has used various forms of RTG on most of its missions to produce power, heat or both. As *Curiosity* was so advanced, the rover needed more power than could be derived from solar cells, and also needed to be weather- and dust-independent. Recall *The Martian* protagonist Mark Watney's daily struggle to clear his solar cells from the thick, caked-on Martian dust that accumulated each night – without upper arm strength or Matt Damon there to help, poor *Curiosity* would be stuck in place. (Watney also digs up and uses the RTG his mission carried to Mars as a source of heat to stop himself from freezing.)

This weather independence also means that *Curiosity* can roam wherever it likes, rather than being limited to regions with long hours of sunlight or having to power down during the darker winter months (as its predecessors did). However, there are some major downsides, mostly linked to longevity.

'There's a bit of irony involved in that *Spirit* and *Opportunity* had solar panels because they originally had 90-day missions,' explains Ashwin Vasavada, NASA's current project scientist for the *Curiosity* rover. 'The thinking back in 2000-ish when they were designed was that the constant rain of dust from the sky would quickly block out the solar panels from the Sun – but for 90 days you're fine. The unexpected thing was that their solar panels kept getting cleared off by wind. They were at particularly windy sites, so that was fortuitous, but you can't count on that. In designing a rover that's intended to last for years from the start, such as *Curiosity*, we chose the radioisotope power supply. That has worked, but now has the disadvantage that you have a known lifetime. It's a long lifetime, but you can't get around it.'

Curiosity has been trundling around on the surface of Mars since August 2012 and its primary mission was intended to last for two years. The rover has evidently long outlived that estimate, and is expected to have a few years still left to go (based on its power supply). However, every day brings new challenges for the rover, including rocky terrain, dust, adverse weather effects, slowly degenerating hardware and more. *Curiosity*'s wheels are damaged from spending so long pushing through Mars's regolith and picking its way around rocks on the planet's surface, but are currently still working (although we've been more cautious since noticing this degradation).

'We think after about eight years on the surface, so around mid-2020, we'll have to slow down quite a bit compared to what we've been doing and spend more time recharging the battery, and things like that,' adds Vasavada. 'But around 8–10 years after landing you can also expect a lot of electronic and mechanical things to start happening, so that's the harder part to predict. For the energy, we can basically plot a curve that we follow exactly until we know how it degrades. For everything else, there's luck involved in, say, not breaking a particularly important part – if a joint on the arm broke, for example, or the wheels were damaged, and so on, we'd lose a major capability.'

Contamination

Every time we venture to Mars, we're careful not to contaminate the Martian surface with any hitchhiking microbes from Earth. This sounds vaguely sci-fi esque, but it's far from it. This is in accordance with the wonderfully named 1967 Outer Space Treaty, which stresses the importance of protecting against two different types of contamination: back (bringing alien matter back to Earth) and forward (carrying Earth matter to another planet).

Contamination is a key consideration for any mission to Mars. We sterilise our spacecraft and carefully consider our landing sites so that we don't contaminate the planet before we figure out whether there is (or was ever) native life there.

However, there's a reasonable chance that terrestrial matter may have made it to Mars. We know that lots of material has been readily exchanged between Mars and Earth in the past (especially during the Late Heavy Bombardment), and we have sent various pieces of equipment there that were not fully sterilised. Luckily, if we have already contaminated Mars it may not be the end of the world: no terrestrial life form would be able to thrive on Martian soil, so the damage may be minimal.

From a selfish perspective, we want to make absolutely, completely sure that any results we get from our experiments aren't false positives. If we went to Mars and found signs of life, we would need to know that we didn't take it with us in the first place.

'You don't want to screw up your measurements by detecting your own life,' says Vasavada. 'When you actually want to detect life, like Viking did, you have to go all the way. They put Viking landers into ovens and baked them – it's amazing that they were able to do that! It's actually gotten harder; when you have a bunch of analogue electronics it's easier to do that than today, when you have all the digital microprocessors and everything else. We've lost the ability to put entire spacecraft into ovens. Fortunately we haven't had to yet as we haven't yet sent the next life detection mission – *Curiosity* is a habitability mission, so we just had to clean ourselves down to a level of contamination that would likely die off pretty quickly in the current harsh environment on Mars.'

Such contamination levels also dictate where our rovers are allowed to go. Because *Curiosity* is only rigorously sterilised on the outside – its 'insides' are comparatively dirty as they are locked within the rover and isolated from the Martian environment, so did not undergo such strict cleaning – there are regions of the surface that the rover is barred from approaching. These are so-called 'Special Regions', designated by the Committee on Space Research (COSPAR).

'Special Regions are places where, to our best understanding, there could still be potential for liquid water today,' says Vasavada. 'That'd be places like the polar regions, where there might be melting or we could induce melting by bringing [a spacecraft] there, or places where there's ground ice, for example.'

In other words, Special Regions are prime real estate for finding Martian life. If we introduced terrestrial life in such regions, it'd have a higher probability of growing and proliferating. These special regions block out both of Mars's poles, with the only 'safe' region existing as a rough central band wrapping around the planet's equator and extending between 0 and 30°/-30° in latitude. Areas between these latitudes and the polar regions are evaluated on an individual basis, and additional patches become protected elsewhere as and when they become interesting (such as the location of the recently characterised RSLs).

As our missions become increasingly sophisticated this may become an insurmountable hurdle, especially if we're intending to send humans. By its very definition, it's simply not possible to sterilise a manned mission to Mars. As well as obviously being alive themselves, humans ooze life and organic matter by breathing, coughing, shedding skin and hair and nails and sweat and other bodily fluids. The chances of us contaminating the planet with terrestrial material are much higher – perhaps even unavoidable. We could lower this risk by constraining humans to certain zones and employing a multi-airlock sterilisation process, but this would increase the complexity of our mission plans and may be hard to implement in the time between landing and setting up a liveable habitat, possibly rendering such measures unnecessary.

Some believe all of this to be pointless on a higher level. To believe that contamination is an issue, one must first believe that we have a responsibility to keep other planets in their 'natural' states, and that our interfering needs justifying (and not all scientists share this view). As mentioned previously, others simply believe it inconceivable that Mars has not been contaminated already. 'If Earth microorganisms can thrive on Mars, they almost certainly already do; and if they cannot, the transfer of Earth life to Mars should be of no concern, as it would simply not survive,' wrote scientists Alberto Fairén and Dirk Schulze-Makuch in a 2013 *Nature Geoscience* article. 'We cannot see how our current program of Mars exploration might pose any real threat to a possible Martian biosphere. The protocols and policies of planetary protection are unnecessarily restricting Mars exploration.'

From a scientific perspective, it may make sense to tread cautiously and explore the planet in as 'biologically reversible' a way as possible. 'Until we know the nature of life on Mars and its relationship – if any – to life on Earth, we must explore Mars in a way that keeps our options open with respect to future life,' wrote Chris McKay, a planetary scientist at NASA Ames Research Center in California, US, in 2009*. 'Mars may well be our first step out into the biological universe. It is a step we should take carefully.'

More astronauts, more problems

Robots are one thing – humans are quite another. As soon as humans enter the mix, the level of complexity and difficulty skyrockets.

Manned missions to Mars will generally be heavier than any of the robotic probes we've sent to the planet in the past, making launch, propulsion and landing a whole lot trickier. Such missions will need to carry enough fuel to get there and back, extra shielding, oxygen and supplies to last multiple years, equipment, scientific instruments, more sophisticated landing gear, life-support systems, the astronauts themselves and more. The sheer length and cumulative weight of such a

---

* In the book *Exploring the Origin, Extent, and Future of Life*, edited by Constance Bertka of the American Association for the Advancement of Science (AAAS).

trip strengthens the call for advanced propulsion; we manage weight limitations just fine for robotic missions, but that's because they are sent into space knowing full well they'll never be coming back and because they weigh less in the first place (although they are still massive).

Manned missions need to develop and deploy reliable and long-lasting life support, communication and navigation systems that we can readily repair, and, crucially, figure out the technology to get back. As rocket scientist and space extraordinaire Wernher von Braun succinctly wrote in a 1965 *Popular Science* article, 'Nobody wants to go on a one-way trip into space'.

The return portion of the trip is a bit of an issue at the moment. No spacecraft would be able to carry enough fuel on the outward trip to later launch from the Martian surface and make the return journey. Additionally, leaving astronauts on Mars to build their own return craft is a dubious idea given the challenges – thick gloves and fumbling fingers, lack of supplies, pervasive dust, weather, the general conditions – so we would want to initially send it from Earth (with the same issue of fuel and the added issue of needing an even bigger rocket to get that one up into space).

Scientists are exploring possible ways of manufacturing fuel on the surface of Mars. We could possibly send a module ahead of any human travellers, land it on Mars and use it to draw the needed components (carbon dioxide, water, maybe methane) from the atmosphere and ground in order to create fuel over the subsequent couple of years. Fuel leakage is also an issue; we would need to keep fuel at cryogenic temperatures to keep it from boiling off, but all of our propellant storage methods are prone to leaking (meaning we'd need to make even more fuel than we believe we'd need). SpaceX, for example, recently announced plans to manufacture and store liquid methane and oxygen propellants on the surface on Mars, by mining Mars's water and drawing carbon dioxide out of the planet's atmosphere.

However challenging, it's likely we'll be able to overcome such technical issues. We know how to launch from Earth, fly to Mars and land on Mars – it's just a case of optimising all of these stages to be suitable for humans.

However, the human body may be a different story entirely. From a known perspective, two of the major issues facing humans on an extended trip in deep space – and later at Mars – are radiation and microgravity.

Radiation

On Earth our substantial atmosphere and magnetic field work together and act like a giant shield, stopping dangerous radiation in its tracks. Mars has neither – and nor does deep space.

There are two main types of harmful cosmic radiation: solar radiation and galactic cosmic rays. Stray particles from both can cause far more harm to the human body than any of the radiative sources we experience on the Earth's surface, and can corrupt or damage space-based electronics.

As the name suggests, solar radiation comes from the Sun. Alongside the ultraviolet radiation it releases, the Sun lets off a stream of charged particles known as the solar wind. This continuous outpouring into space causes bizarre and beautiful phenomena on Earth as it interacts with our atmosphere (aurorae, for example). The Sun's magnetic field is very fluid and, as a rough-and-ready analogy, winds up tighter and tighter over time, tying itself in knots. Occasionally it will twist itself up so tightly that it effectively snaps, releasing a huge burst of radiation in the form of a solar flare or more sizeable 'coronal mass ejection'. These are so dangerous that space-based capsules such as the *International Space Station* (*ISS*) have special insulated rooms that astronauts can shelter in if it looks like a solar ejection is headed their way.

Cosmic rays, on the other hand, are more exotic. They are flung into our Solar System from far beyond our little patch of space. They're mostly made up of intensely energetic protons and atomic nuclei, thought to have been thrown out from violently dying stars, or from superheated material circling around black holes at the hearts of distant galaxies. These rays can be very dangerous when they hit the Earth's atmosphere, as they can cause secondary sprays and showers of particles that filter down to the Earth's surface.

When performing spacewalks within a particularly radiation-dense region of our atmosphere (known as the 'South Atlantic Anomaly' or SAA), some astronauts have reported seeing streaks of light akin to 'shooting stars' moving across their field of view; NASA astronaut Don Pettit described them as 'flashes in [his] eyes, like luminous dancing fairies', which he saw 'even in the dark confines of [his] sleep station, with the droopy eyelids of pending sleep'. These 'fairies' are actually charged particles hitting the retina. In normal orbit an astronaut might experience one or two 'flashing fairies' every 10 minutes, but in the SAA it increases to multiple per minute. The longer a spacecraft is in flight and without protection – on a trip to Mars, for example – the bigger this hindrance becomes.

While both radiation sources are problematic, cosmic rays are particularly worrisome because we cannot effectively predict and shield against them, making them the biggest danger to those aboard the *ISS*. While a stray cosmic ray can cause havoc with space-based electronics, for human exploration it could be a matter of life or death.

Being exposed to such a powerful and dense burst of radiation could considerably increase an astronaut's chance of cancer. This is one of the key hazards for radiation workers on Earth and even for airline pilots, who spend a lot of time at higher altitudes than the average person. During a year on the Earth's surface, galactic rays bathe the average person in radiation totalling just a fraction of a unit known as a millisievert (mSv, a measure of our body's reaction to ionising radiation). Including all other sources of radiation – materials used for nuclear power, radon gas, consumer products and medicines, even rocks and concrete, which are mildly radioactive – you probably experience a few mSv of radiation per year. For context, you'd receive up to 10mSv when going for an abdominal CT scan, so this isn't cause for concern!

Radiation exposure is currently an unavoidable hazard for astronauts. Space agencies across the globe are strict about how much radiation one crew member can soak up during their career. Most agencies – ESA, JAXA, Roscosmos – cap the lifetime exposure for astronauts at around 1,000mSv. NASA limits its astronauts' exposure-induced increased risk of fatal cancer to 3 per cent, which translates to a cumulative lifetime dose of between 450 and 1,500mSv dependent on age and gender.

Being exposed to roughly 1,000mSv of radiation in a short timeframe increases your risk of developing fatal cancer in your lifetime by a significant 5 per cent, so this is no small concern. In the US, for example, radiation workers can only be legally exposed to a maximum of 50mSv annually. In space, the numbers are much higher. Astronauts visiting and spending time on the *ISS* are exposed to double this dose in just half the time – and the environment of deep space is hundreds of times more intense.

Mars Science Laboratory (*Curiosity*'s ride) measured this on its way to Mars, switching on one of its instruments and measuring an effective radiation dose of 1.8mSv per day. At this rate, astronauts spending six months in deep space would soak up roughly 330mSv. This might sound feasible given the limits just mentioned, but consider how long it'd take for an astronaut to visit Mars and return to Earth. If a manned mission to Mars were to launch tomorrow and simply loop around the Red Planet, the on-board crew would need to spend nine months on the way there, a few months hanging around waiting for the right return window and nine months back – say 600 days (likely an underestimate). This would dole out a cumulative mission dose of 1,080mSv … dangerously close to NASA's lifetime limit for radiation-hardy astronauts and above it for most other agencies.

Unfortunately, this doesn't include any previous trips to space – ruling out every astronaut with previous experience – any particularly large bursts of radiation thrown off by an unexpected solar storm or event, any extra time spent hanging around waiting for the best return orbital window, any delays or, crucially, any time spent on the surface of Mars.

There are ways to combat this problem. The primary method is to work on more effective ways to shield any spacecraft carrying intrepid Mars explorers. This is currently done for the *ISS* using passive shields and by packing various materials into the walls of the craft. The craft's walls themselves also act as shields, but are nowhere near thick enough to make a significant difference on their own. As hydrogen is known to be an effective shield, water and food are often used to line the inner walls of a shelter, turning the astronauts' vehicle into a kind of giant cosmic refrigerator. This water lining would unfortunately add weight, but might be necessary. It's also been suggested that common plastics, such as the hydrogen-rich polyethylene used in most plastic packaging, might also do the job well, as might sturdy little structures known as hydrogenated boron nitride nanotubes (nano-sized tubes of boron, nitrogen and carbon filled with hydrogen). A little less appealing is the idea of slowly adding compacted faeces to the walls during the mission, which could strengthen the shielding – and make for a far less pleasant confined trip!

There are other promising technologies, such as magnetic shielding (real-life force fields!), that could help to solve this problem. Scientists are also working on deployable storm shelters, which could give additional protection in the case of an unexpected solar event, and are working hard to understand the erratic behaviour of the Sun, which could indirectly help us to avoid the worst of it. Other than that, the best thing we can do is to simply get to Mars faster.

This issue continues on the Martian surface. We cannot just erect our tents on a flat piece of ground with a nice view, as the radiation environment is intense and dangerous. While lessened, the radiation environment on Mars is still a problem, with around half the radiation levels experienced in interplanetary space. To tackle this, various mission plans propose shielded capsules or futuristic pods (pressurised to deal with the low Martian air pressure and heated to handle the low temperatures). Others suggest digging deep down into the planet's soil, creating little burrows that could then be pressurised and pumped full of breathable air, and adding a rocky layer of natural radiation protection.

This is all part of a strategy known as 'in-situ resource utilisation' (ISRU), which is a fancy name for 'using the stuff we find on Mars in an effective way so we don't have to bring it with us'. This includes using the atmosphere and rock to manufacture fuel, separating the carbon dioxide in Martian air into carbon and oxygen so we can breathe, building nuclear plants or giant solar panels to harness solar energy, and sourcing water from either underground stores of ice or hydrated salts and minerals laced through the soil. We also hope to process Mars's soil so that we can grow food in it – we know that lettuce, cress, rice, soybeans, radishes, potatoes, onions, peas, herbs and certain types of wheat and flax grow just fine in space so might work in Martian gravity, too. This would give a community a long-term future, reducing its reliance on expensive supply shipments from Earth.

Microgravity

Deep space challenges the body and psyche in almost every way. To put it simply, humans are just not very good at being in space.

Build and height

In microgravity our tendons relax and our spine begins to stretch out, adding a few inches to our height. However, even the shortest of astronauts wouldn't welcome this; it's painful, and reverses upon return to Earth. The same effect is seen with the chest and ribcage, giving astronauts a barrel-chested appearance. Our muscles relax, the ribcage expands outwards, and our internal organs experience a weird sensation of weightlessness, as if they're floating upwards inside the body.

Muscles

Bodybuilders beware: microgravity is a direct enemy of muscle growth. Muscles don't have much use in microgravity. Astronauts can get around with very little physical exertion and thus use their muscles far less. This results in pretty serious muscle wastage, or atrophy, over time. All astronauts return to Earth weighing far less than when they departed. Studies have shown that astronauts return to Earth after just 180 days on the *ISS* with similar muscle tone to the average 80 year old, despite being well above the average level of fitness when they left. Astronauts on the *ISS* try to fight this via a strict high-resistance exercise regime, strapping themselves into resistance machines and working out for at least a couple of hours per day.

Bones

Just as high-resistance exercise can build up stronger, denser bones and stave off brittle-bone conditions such as osteoporosis, lack of gravity can do the opposite, causing bones to weaken and become more fragile and less dense (a condition known as 'spaceflight osteopenia'). Astronauts lose bone mass at a rate of roughly 1–2 per cent every month; the average elderly person or post-menopausal woman loses this amount *annually*.

Blood and urine

Our weakening bones have another less-than-ideal side effect. As calcium is removed from the skeleton, it's added to the blood. Excess calcium in the blood is one of the key causes of kidney stones; others include changes in diet and urination (acidity, volume), dehydration, increased consumption of salt and the quality of water intake. Sadly, all of these risk factors combine in spaceflight, leaving astronauts highly at risk of developing kidney stones. Luckily there has only been one in-flight incidence of this to date, when cosmonaut Anatoly Berezovoy reportedly suffered a crippling pain in his side while aboard the *Salyut* 7 space station in 1982. Suspecting appendicitis, the crew began preparing for an emergency evacuation – until Berezovoy passed a small kidney stone and promptly recovered. This might seem like a small problem, but it'd be very difficult to treat in space if there were complications. It's easier to prevent than to treat, so officials are looking at possible ways to reduce this risk via increased hydration, nutrition and supplements*.

Heart and circulation

On Earth our hearts need to be relatively large in order to effectively fight gravity, pushing and pulling and pumping blood towards and away from our extremities. In microgravity the heart continues to do what it's always done, but the distribution of blood and fluid within the body is different; as a result, fluid pools in the upper parts of the body, and we experience a rise in heart rate and blood pressure. Over time, the heart stops working so hard and begins to lose muscle mass, shrinking and becoming more spherical. Our circulatory system also struggles to adapt in space; the change in gravity affects the shape of our red blood cells, making them more spherical and allowing fewer to fit in the same blood volume. Our body therefore stops producing them, leaving many astronauts with a form of anaemia (iron deficiency) in just a few days. There may be even greater issues here that we're currently unaware of; there are warning signs that astronauts who travel to deep space may have a far greater risk of developing cardiovascular diseases and complications. A 2016 study found that 43 per cent of deceased Apollo astronauts died from a cardiovascular problem – a percentage 'four to five times higher than non-flight astronauts and astronauts who have [only] travelled in low Earth orbit'.

---

* The increased calcium levels in astronauts' bloodstreams led to complications with a urine processor assembly (UPA, designed to recover water from urine) on the *ISS* in 2009. It worked well on Earth, but the increased calcium levels and decreased volume of astronaut urine ended up hindering the filtration process. The calcium precipitated on the equipment and clogged it! This flags up a key issue with manned spaceflight – to meet current mission timelines we will need to send a crew to Mars with technology that we haven't tested in a genuine Mars environment. Any new and unforeseen issues could throw a major spanner in the works. Due to Mars's distance, we may not be able to replace or repair any damaged equipment for some time.

Brain and cognition

The brain appears to work slightly differently in space. Basic tasks like coordination, critical thinking and attention span appear to be impaired and slightly slowed. This could possibly be due to a curtailed blood flow caused by microgravity, but we're unsure.

Vision

Space may be beautiful, but it's certainly not easy on the eye. Many astronauts returning from the *ISS* experience severe problems with their eyesight. Such issues were initially thought to be temporary, as some vision problems have shown to clear up in the months and years after an astronaut has landed. However, post-flight studies performed on 300 American astronauts since 1989 – astronauts flying on short two-week missions or longer five-to-six-month missions – showed that 29 per cent of the former and a significant 60 per cent of the latter fliers experienced impairment and degradation of their visual acuity (keenness) that didn't improve post-landing. As recently as a couple of years ago this problem was labelled a potential 'showstopper' by NASA's Human Research Program. A crew that were slowly going blind would be a huge problem! We're still unsure exactly what causes this problem, but it may be due to the weightlessness an astronaut experiences, which causes fluid to redistribute itself throughout the body. On Earth gravity keeps our fluids towards the lower parts of the body, but in microgravity it quickly rushes up towards the head, making it feel puffy, swollen and congested, and increasing the pressure on our optic nerves and eyeballs.

Appetite

As well as draining fluid towards the top of the body, resulting in thinner 'bird legs', the increased puffiness in an astronaut's head can cause symptoms similar to a head cold. It dulls our sense of taste and smell, making one less interested in eating and less able to enjoy it. This may seem like a blessing in disguise given the quality of the in-flight menu, but astronauts often show a tendency to under-eat because of this, which is a big problem. They need to pack in the nutrients and keep their weight up – especially as they're working abnormally hard during their long office hours and exercising for hours each day. Scientists are constantly developing new and diverse dishes for astronauts, from cookies to spaghetti to kimchi, to keep them interested in their meals.

Balance and coordination
Astronauts often experience disorientation, discomfort and nausea, especially during takeoff and in the earlier stages of their missions. This is known as 'space adaptation syndrome', and is essentially a form of motion sickness triggered by changing g-forces and vestibular stimuli (sensory input related to balance/orientation). It affects our ability to locate which way is 'up', severely impacting our spatial awareness and balance. Astronauts find this troublesome for quite a while after returning home, stumbling around, sitting and standing abruptly or awkwardly, and struggling to coordinate their movements and steady their gaze. Our ears are especially affected, as they help us to maintain balance and determine our orientation (if you've ever had an ear infection, you'll be painfully aware of this). In microgravity, directions are muddled and astronauts often experience odd and rapid flips in their perceived orientation, and frequently feel as if they're 'upside down'. This becomes particularly worrisome when considering an astronaut's decreased bone density and higher risk of fractures, but would hopefully improve once astronauts reached the Red Planet.

Rest and recovery
The unusually high levels of stress, anxiety, change and disrupted sleeping patterns experienced by astronauts weakens their immune systems, and can lead to exhaustion and fatigue. Some suffer from chronic sleep deprivation. In fact, up to half of some *ISS* crews rely on sleeping pills to force themselves to get the rest they need to fulfil their duties, but even so the

average astronaut sleeps for two fewer hours per night than they did on Earth.

Body clock

Our body's 'inner clock' (circadian rhythm) relies on external stimuli, such as the fluctuating levels of sunlight we receive during night, day, dawn and dusk, to keep itself running correctly, ensuring we're tired at night and refreshed in the morning. Biological processes on Earth appear to work on a natural timescale of 24 hours (the so-called circadian rhythm, aligned to the length of one Earth day). According to NASA, 'almost every physiological and psychological system evaluated thus far' has been shown to be affected by circadian timing. On the *ISS*, astronauts experience a completely different reality; the station orbits Earth once every 90 minutes and so they experience 16 sunrises and sunsets in a single 24-hour period, meaning that a 'day' essentially lasts for 45 minutes. Astronauts have to work hard to align their body clock with their new environment, a process that sometimes begins prior to launch. Mars has many properties that might affect a human's circadian rhythm, including a slightly longer day, different levels of sunlight and a yellow-brown surface (which in turn colours the air and sky a salmon-pink hue). Our research into circadian rhythms doesn't use a Mars-like environment as the norm, so we don't yet know how our bodies will react to Martian stimuli*.

Sexual activity

Even if you happened to take your partner with you, as one organisation has suggested, you'd almost certainly be unable to indulge in anything intimate until we know more about

---

* Exposure to light seems to dominate the human circadian rhythm. It affects our mental acuity, hormone levels and body temperature throughout the day. For example, there's a measurable slump in our body temperature and associated mental alertness in the early afternoon (so the urge to devour a 3pm coffee and biscuit pick-me-up is mere biology!).

the logistics of space sex (and yes, that is a real, albeit theoretical, research topic). The changes to fluid and blood distribution throughout the body would likely make it difficult for men to 'perform'. If you wanted to try anyway your vital signs would be monitored throughout, giving you little to no privacy – and definitely ruining the mood.

Mental health

We've long been aware of the problems that crop up if you coop people up in a small space for a long time. Astronauts heading to the *ISS* often struggle with the lack of home comforts, physical affection, emotional intimacy, variety (in terms of company, food, routine and more) and even a form of separation anxiety caused by being away from Earth for so long. Many astronauts have experienced something known as the 'overview effect', where seeing Earth from space caused them to instantly realise how small and devastatingly fragile our planet really is. These problems would be amplified on a trip to Mars as any crew won't be able to return home easily (if at all). The time delay between Earth and Mars – which ranges from 4 to 24 minutes depending on the orbital positions of the two planets – also means that a crew would have the added stress and pressure of needing to be largely autonomous, especially in emergency situations, and rules out the possibility of smooth communication with Earth, making it difficult to track developing issues or to conduct therapy sessions. NASA has even pondered the possibility that a crew might sabotage a long-term mission on Mars due to the various psychosocial issues that might arise (smaller 'mutinies' have occurred previously such as the *Skylab 4* mutiny of 1973, where one crew shut off their radio communication with NASA for a whole day after complaining of exhaustion and frustration at their hectic schedule).

Social health

Imagine being shut in a small space with a group of people you don't know all that well for nine months. Add an extended stay on the surface of Mars, say for 12 months, and a nine-month return trip – between two and three years in total, and that may be an underestimate. Each of your crewmates will have their own little quirks and idiosyncrasies; Steve might have an annoying laugh, Anna might chew with her mouth open or clear her throat every couple of minutes, John might speak inconceivably slowly. After three years, these harmless little quirks may not seem so little or harmless … and there's no escape. Individuals who have endured long expeditions on Earth – including Frederick Cook, an American doctor who claimed to have been the first to reach the North Pole in 1908 – have complained of monotony, morale slumps, emotional exhaustion and 'spells of indifference' akin to depression. 'Around the tables … men are sitting about sad and dejected, lost in dreams of melancholy,' Cook wrote in his diary. 'We are at this moment as tired of each other's company as we are of the cold monotony of the black night and of the unpalatable sameness of our food.' Russian cosmonaut Valery Ryumin was a little more blunt in his diary. 'All the conditions necessary for murder are met if you shut two men in a cabin measuring 18 feet by 20 and leave them together for two months,' he wrote.

Personality

This highlights another somewhat sensitive problem: the inevitable clash of different cultures and personalities. Some people just don't gel well with others. Astronauts may be highly trained professionals, but they're still human. How do you ensure that a crew is well-suited in a social sense for such a long period of time? For example, what if one crew member is a tenacious lover of the Rolling Stones while the other is a hardcore fan of the Beatles? On a more serious note, what if one is highly religious while another is a staunch atheist, or vocally right wing versus left wing? All it takes is for one person to also become outspoken or abrasive – qualities that are more likely to develop or come out over time due to social friction – and it could sour interpersonal relationships within a crew. One Earth-based space simulation, named Sphinx-99, had serious consequences in this area. The mission took place in Moscow in 1999, and had a crew of six men – four Russian, one Japanese and one Austrian – and one woman, Canadian Judith Lapierre. After less than a month, Lapierre was sexually assaulted by the Russian commander, who pulled her out of sight of the surveillance cameras to kiss and grope her. The Russian 'ground crew's' response? Lapierre 'ruined the mission, the atmosphere, by refusing to be kissed'.

It's incredibly easy to become indignant here, but on a pragmatic level this strongly emphasises the importance of cultural sensitivity training for all crews, and a well-rounded selection process to head off issues like this before they crop up. Some have suggested single-sex or all-female crews, which may alleviate some issues – including making a mission lighter and cheaper, as women generally weigh and eat less, and avoiding vision issues, as women don't appear to suffer as much as men* – but would be unlikely to eliminate social problems altogether. Some space agencies state that female astronauts are better at building relationships and communicating with their crewmates, and adapt better to spending time alone (although NASA reports no gender difference).

All of these problems are great for anyone spending time in space, but become greater as we add increasing amounts of time on the Martian surface. Mars's gravity is just over a third (38 per cent) of Earth's, so gravity-related symptoms may persist even once a crew had reached their destination. We're not yet sure how to mitigate or avoid some of these effects – or if the human body is ready to undertake such a challenge.

Ethics and funding

One of the most pressing limitations in getting to Mars may be something far more mundane than space science. It might actually be a case of political science – of money, management and politics.

* This may well be due to there being a far lower sample size of female versus male astronauts, so the correlation is still unclear.

Some have indicated that public support is a key factor in setting foot on Mars, and that without such support any programme aiming to do so will fail (this only serves to highlight the real significance of previous chapters and is a common theme in the space science arena). Any organisation with Mars in its sights will need to secure reliable and continued funding and support for an expensive science programme that needs to span many decades in order to succeed. Funding for a couple of years is no good: any Mars programme needs funding for over a decade to simply get off the ground.

Taking NASA as an example, such continued funding looks unlikely. The agency is constantly fighting budget cuts and its financial future is uncertain from one year to the next – a disaster for long-term science. Even if supplied with an injection of funding, this may be allocated to a different department or project (for example, some have objected to NASA's *Space Launch System* rocket receiving such huge amounts of money, as it effectively saps areas of EDL research very pertinent to Mars). As a result, it's likely that future missions may aim to be globally collaborative, involving multiple space agencies and organisations.

While this may not – and hopefully will not – be a showstopper, it's certainly on many minds. This is perhaps why multiple private organisations have entered the game with their own proposals for reaching Mars. While it is not guaranteed that such private endeavours could reach Mars quicker, more reliably or more cost-effectively than any government-led plan, the chances of avoiding bureaucracy and fluctuating budgets may be higher.

Issues may arise, however, when public and private organisations differ on ethics. For example, private company Mars One has proposed a one-way trip to Mars, on the basis that it would be easier, cheaper and we could therefore implement it sooner. However, government agencies would struggle to convince the public that such a plan was appropriate. Sending humans into space to never return home and ostensibly die on another planet is a never-before-tackled ethical grey area – and a really hard sell.

Previous missions to Mars have only encountered the issue of ethics in relation to contamination, but humans add a whole host of other quandaries. We are also new to the interplanetary exploration game and are painfully aware that we don't know what we don't know. Pretty much every space agency with designs on Mars has analysed the risks involved and found them to be great – and, in some cases, maybe even insurmountable.

We may be sending humans to Mars knowing that they will never return, or that they are highly likely to injure themselves fatally or die young due to radiation exposure. If Mars colonisers become ill or show signs of sickness, we may not be entirely sure whether they are allowed to come back to Earth; do we know if it's the common cold, or some as-yet-undetected Martian pathogen that could trigger a global catastrophe here on Earth? We may even need to refuse calls for help from a Martian colony if we want to stay on target and in budget with future plans; this issue was raised in *The Martian*, where stranded astronaut Mark Watney was only rescued due to his former crewmates deciding to rebel against NASA instructions and mutiny, returning to Mars to collect him. It may seem cruel or heartless, but it is necessary to temper our emotions with pragmatism if we want to set up and maintain a long-term presence on Mars. It's almost inevitable that people will die in our attempt to get there, both in the short and long term.

Is it ethical to do this? Is it really worth sending humans to Mars at such risk?

'What makes risk ethical? Historically it has been one thing: consent,' wrote Laurie Zoloth, professor of medical ethics and humanities at Northwestern University in Chicago, US, in *Cosmos* magazine in 2015. Zoloth pointed out that 'the ethical considerations change if we think of the crew as military personnel' or as 'pioneers'. 'We expect soldiers to face considerable risk,' she wrote. What makes astronauts any different?

This opinion has been echoed by SpaceX magnate Elon Musk, who is aiming to send humans to Mars in the coming decade. He has said that 'people will probably die – and they'll know that'. As long as appropriate measures are taken to protect astronauts, and their contribution and sacrifice is recognised, informed consent may lower or negate many of the ethical concerns involved in going to Mars.

Ethics aside, the question of whether it is worth sending humans is easily answered. It is hard to overestimate the effect that sending humans to Mars would have on our understanding of the planet. The rovers, while amazingly successful, trundle around very slowly and carefully consider every single move they make. After all, one particularly jagged rock under their wheel could leave them stuck (although this would no longer be a big worry if we were there to repair them). *Spirit* and *Opportunity* landed on Mars in 2004. *Opportunity* has covered 43km (26.8 miles) in the past 12 or so years and *Spirit* clocked up 7.7km (4.8 miles) during its five years of mobility. In the past four years, *Curiosity* has managed to travel nearly 14km (8.7 miles). That's a combined total of around 65km (40 miles) in more than two decades of combined operation. How far did you drive last week in your car?

A team of astronauts could do more science on Mars before dinner than a rover could in months. Humans can repair things, think on their feet and adapt, communicate issues and head them off before they worsen, innovate, gain a larger picture of their surroundings and spot erroneous or anomalous details, and understand more in order to do more relevant and insightful science. When *Curiosity* releases a new batch of images, it's common for people to joke that if the rover just swung its 'eyes' a little to the left, we'd see a family of aliens sunbathing or enjoying a leisurely picnic – obviously not, but the point stands that the rover simply 'sees' what it is told to see.

'Although amazing and productive, the amount of time and effort we have so far been able to spend studying Mars is minuscule compared to the time and effort spent investigating our own home planet here,' says Penelope Boston, director of the NASA Astrobiology Institute in California, US. 'So, it's no wonder that we are faced with so very many questions to be investigated over the years to come. Mars is a fabulously complex world with the same span of history that Earth has, but many different circumstances. It is an exciting time to be alive and to see our species begin to take its baby steps towards the rest of our Solar System.'

CHAPTER TWELVE

The Future of the Red Planet

Cold, airless, dusty, dangerous, distant and expensive. Who would want to go to Mars?

Apparently, pretty much everyone! The coming decade is full of proposed, planned and scheduled launches to the Red Planet from agencies and companies all across the globe.

ESA and Roscosmos will soon launch the sibling to *ExoMars 2016*, *ExoMars 2020*. The lander–rover combination is expected to land on Mars in 2021 and will begin to hunt for possible signs of life; the solar-powered rover will explore the surface and hunt for signs of biological activity, while the landing platform will study the planet's climate, surface, internal structure, rotation and orbit, the amount of water vapour in its atmosphere and more.

NASA intends to launch a lander named *InSight* (Interior Exploration using Seismic Investigations, Geodesy and Heat Transport), which aims to study the inner workings of Mars's deep interior. The original plan was to launch *InSight* in 2016, but delays have shifted it to 2018. The agency is also working on its *Mars 2020* rover, which will be a successor – or maybe even colleague – of *Curiosity*. Mars 2020 will gather regolith samples and hunt for signs of a biosphere (simply put, signs of any kind of ecosystem, including interactions between life and its environment either in the past or present) and further *Curiosity*'s quest by widening the focus from habitability to life itself. Excitingly, the rover will carry a microphone with it to the Red Planet, allowing us to 'hear' Mars for the very first time.

India has another mission in the pipeline, *Mangalyaan 2*, scheduled for 2020, which will comprise an orbiter and possibly a lander–rover component (partly in collaboration with France's National Centre for Space Studies, CNES).

July 2020 will see a busy time for our skies; alongside *Mangalyaan 2*, *ExoMars 2020* and *Mars 2020*, the United Arab Emirates is aiming to launch its *Mars Hope* orbiter (Emirates Mars Mission) and China is intending to launch its 2020 Chinese Mars Mission, an orbiter–lander–rover triple whammy.

There are additional proposals to these missions, but they remain unconfirmed. However, perhaps more exciting than these robotic plans are the proposals for manned missions to Mars, with the aim of establishing a long-term human community on the planet in the next two decades or so. This is slightly reminiscent of science fiction's imaginings of the Red Planet and does sound very futuristic. However, while some plans are more ambitious than they are feasible, many others are far from extraterrestrial pipe dreams.

NASA is aiming to set human foot on Mars in the 2030s as part of its 'Journey to Mars' programme. It is developing and building what will be the world's largest and most powerful rocket in the form of the *Space Launch System* (*SLS*). The *SLS* will be responsible for delivering *Orion*, NASA's proposed Mars-bound spacecraft, into space so it can begin its journey to Mars. The rocket will be capable of hauling 130 tonnes (287,000lbs) into space and is primarily intended to help payloads get beyond Earth orbit.

The *SLS* is an expendable rocket. It'll carry its payload into space, stop just shy of entering orbit around the Earth, release its cargo and fatally tumble back down to Earth. Successive *SLS* launches will carry missions destined for locations further and further out into the Solar System; while the first few will be unmanned, in the 2020s NASA plans to send four-person crews to orbit the Moon and a nearby asteroid (which will be captured and dragged into orbit around the Moon). Each launch will test the machinery and technologies we'll use to get to Mars, and allow us to adapt the SLS boosters and architecture as needed. If successful, NASA then hopes to send humans to Mars aboard *Orion*, launched into space via the *SLS*, in the 2030s.

The *SLS* has proven to be controversial. Many have criticised the rocket as being too expensive and for sucking funding away from other parts of the Mars programme, instead calling for more innovative methods of propulsion and reusable launch systems. NASA has defended the expendability of the *SLS*, stating that it removes the cost involved in recovering, transporting and fixing crashed rocket boosters, that the design itself requires disposable parts and that it's more cost-efficient to let the hardware go.

NASA has pinned down some key timelines, but is still considering the best steps forward and anticipating an international and globally collaborative approach. Its current plan is a three-step one: first, scientists will glean as much relevant information as possible from our *ISS* experiments on human health, ISRU, life support, communications and more ('Earth Reliant' exploration), then send research crews increasingly further out into deep space, primarily into the region between Earth and the Moon (the 'Proving Ground' stage, which includes capturing an asteroid circa 2020, advancing our resupply and habitation options, and testing the *SLS/Orion* duo) and then work on the showstopper ('Earth Independence') – sending humans into low-Mars orbit or to one of the Martian moons and then down on to the surface of the planet.

'While far away, Mars is a goal within our reach,' NASA announced in its 2015 'Journey to Mars' report, in which it outlined the details of its plan. 'We are closer to sending humans to Mars than at any point in NASA's history.'

The European Space Agency has considered sending humans to Mars in the past via its Aurora programme, but changes in funding and support have put these plans on hold. ESA has conducted collaborative space simulation tests here on Earth to see how crew members would cope on a trip to Mars, notably from 2007 to 2011 in the form of the very successful Mars-500 simulation. For now, ESA is focusing on robotic exploration of the Red Planet. Similarly, Russia's Roscosmos and the Chinese CNSA are focusing on the 2040s to 2060s to send humans to Mars, but have not yet pinned down any specifics.

Going private

These kinds of roadblocks may have spurred private organisations to join the Red Planet race – it's unclear if a private timeline would be shorter than a governmental one, but the lack of bureaucracy and public responsibility might accelerate the process.

The most notable entry is Elon Musk's SpaceX, a company that has dominated the discussion around private missions to Mars in the past few years. Various private agencies and organisations have spoken of going to Mars but, SpaceX aside, there have been no confirmed proposals. SpaceX is working on its version of the *SLS* in the form of the colossal *Falcon Heavy*, which would have 'the ability to lift into orbit over 54 tonnes (119,000lbs) – a mass equivalent to a 737 jetliner loaded with passengers, crew, luggage and fuel'. The company has worked closely with NASA in the past, and intends to rely on the agency for technical and operational support.

Alongside its *Falcon Heavy* heavy-lift rocket, SpaceX has plans to send humans to Mars in the near future. However, Musk is dreaming bigger than a mere fly-by or short-lived jaunt on Martian soil. While NASA plans to take small steps towards establishing a human presence on the planet, Musk has announced accelerated plans to build a huge human settlement on Mars in the 2020s and 2030s, beginning with multiple unmanned launches in 2018 and every couple of years after that (carrying supplies, equipment and rovers aboard a capsule named *Red Dragon* – creating a 'regular cargo route to Mars every 26 months, like a train leaving the station', Musk has said). SpaceX then aims to send humans to land on Mars in 2025 aboard a craft known as the *Interplanetary Transport System* (formerly the *Mars Colonial Transporter* – this craft may undertake a test launch a couple of years prior to an attempted manned launch, in 2022).

'I do want to emphasise this is not about sending a few people to Mars,' Musk said to the *Washington Post* in June 2016. 'It's about having an architecture that would enable the creation of a self-sustaining city on Mars with the objective of being a multi-planet species and a true space-faring civilisation.'

Although Musk is seen as highly capable, he has been famously optimistic in his plans so far, fuelled by his obvious excitement and impatience to get to Mars; SpaceX mission plans and announcements have been repeatedly delayed, and Musk has admitted that a fair amount of luck will be needed in order to get humans on the surface of Mars by 2025. At the time of writing Musk had just announced the preliminary details of SpaceX's plan at the International Astronautical Congress in Mexico (September 2016), confirming the mission timeline listed above and adding more detail about fuel production, launch architecture, cost, crew size and more, claiming the ability to create a 'fully self-sustaining civilisation on Mars' within the next century. While incredibly inspiring and technically feasible, his plan quickly garnered some familiar criticism – namely that Musk's timelines are very ambitious (as he admitted himself during the announcement) and massively expensive, and that there are many hurdles yet to be addressed and overcome.

Other private companies have put forward tentative plans for a Mars mission architecture, or entered the Mars space race in different ways. Boeing, for example, is working with NASA to transport astronauts to and from the *ISS*, and proposed its conceptual 'Affordable Mars Mission Design' in 2014. Orbital ATK and Richard Branson (Virgin Galactic) have also expressed interest in Mars, though the spaceflight contributions from the former come primarily through its work on NASA's *SLS* (for which they are providing the boosters) and low-orbit commercial ventures around Earth (including shuttling cargo to the *ISS*), and from the latter via its quest to make spaceflight as routine as driving a car (the birth of space tourism!). Branson has spoken of his desire and determination to '[be] a part of starting a population on Mars' within his lifetime, but has no firm plans.

A few years ago two different private agencies, Mars One and the Inspiration Mars Foundation, prominently challenged NASA's aim of sending humans to Mars by the mid-2030s, proclaiming they could do it far more quickly.

The Inspiration Mars Foundation was founded by entrepreneur Dennis Tito in 2013. It was immediately viewed as overambitious; the organisation was aiming for a manned mission to Mars in 2018, but no landing – a fly-by. While this would still be difficult, with no landing, rendezvous, docking or other complex manoeuvres it was potentially feasible on such a short timescale. However, the foundation required investment from NASA, SpaceX or another source in order to make its plans a reality, and both were unconvinced by Tito and busy with their own endeavours. The Inspiration Mars website is now offline and the company's future is dubious at best.

Mars One, a company founded by Dutch entrepreneur Bas Lansdorp, has proposed a one-way trip to Mars that would land on the planet's surface in the 2020s. The idea of a one-way trip may be ethically controversial, but it would halve the deep-space radiation risk and thus might be a quicker way to reach the Red Planet. According to Mars One's plans – which have been progressively delayed and altered over past years – the company aims to send an unmanned 'demonstration mission' and communications satellite to Mars in 2020, a rover and second comms satellite in 2022, six cargo missions in 2024 and one-way crews of four to the Red Planet every 26 months beginning in 2026.

Scientists and engineers worldwide have been openly critical of Mars One. For one, the company appeared to have no feasible plan for funding – one of its proposed sources of money was to establish a Mars-based reality TV programme, where interested Earthlings could tune in to a kind of Martian *Big Brother.* Lansdorp has continually dodged questions about funding, timelines, technology, crew – essentially all elements of his plan – and has repeatedly delayed his deadlines. Many remain cynical about whether the required technology for Lansdorp's plans even exists yet. An independent scientific study performed in 2015 by scientists at the Massachusetts Institute of Technology (MIT) investigated Mars One's plans in more detail – with a damning result. 'The Mars One mission plan, as publicly described, is not feasible,' they deduced.

There are also queries about ethics. For example, NASA fully believes that it won't be ready to tackle Mars One's dream of sending humans to Mars until the 2030s, and the agency has been a key player in the Mars game for decades. If NASA is currently unable to confidently send astronauts before then, it seems surprising – or unlikely – that an underfunded private organisation could safely do the job over a decade sooner. Is this a case of being overly cautious on NASA's part (after all, the agency does have to navigate a lot more paperwork and red tape), or impulsive on the side of companies like Mars One? How do these private plans have the technology to do this if even an agency the size of NASA is still working on it? In 2015, NASA administrator Charles Bolden phrased these concerns rather more bluntly: 'No commercial company without the support of NASA and the government is going to get to Mars.'

Despite all the dangers and the criticisms of Mars One, and despite it being one-way, wannabe Mars explorers flocked in their thousands to volunteer for the trip (for which they had to pay a registration fee of up to US$75). It seems that for many, the idea of setting foot on Mars is enough to outweigh any possible dangers or misgivings. Mars One initially claimed to receive over 200,000 applications for its proposal to colonise Mars … but then admitted that only 4,227 candidates were willing to pay the admin fee. This number has since been whittled down to just 100 potential astronauts, but even these willing, and paying, volunteers aren't completely convinced.

One of the 100 candidates, Joseph Roche of Trinity College Dublin, Ireland, expressed his concerns in a 2015 interview with *Matter.* He claimed not to have met anyone from the organisation in person, to have been chosen purely based on the amount of his donation and to have only been vetted via a 10-minute Skype call. 'That means all the info they have collected on me is a crap video I made, an application form that I filled out with mostly one-word answers … and then a 10-minute Skype interview,' Roche said. 'That is just not enough information to make a judgment on someone about anything.'

The Mars One critics are almost certainly correct in their cynicism; every one of Lansdorp's grand plans has been delayed by several years. Lansdorp had declared a budget of just US$6 billion, a tiny, tiny fraction of NASA's estimate, but in 2016 admitted that even this amount was 'currently not in place'. As it stands, Mars One appears to be more of an overambitious dream than a viable mission and so should be taken with an astronomical pinch of salt.

Terraforming: The birth of Earth 2.0

Despite its shortcomings, Mars bears more similarities to our home planet than any other world in the Solar System. It is rocky, we recognise much of its geology, it has polar caps and a moon or two, its axial tilt and length of day are pretty close to Earth's, its size and gravity aren't so different that the planet is an impossible prospect for visitation, its surface temperatures creep into a manageable human range and the planet's rock, soil and atmosphere seem to contain many familiar elements. We see a lot of ourselves in Mars.

There are various sites we use on Earth – so-called Mars analogue locations – to prepare ourselves for our missions there. Compare a photo of a canyon in Utah, US, or a panorama from the Chilean Atacama Desert to one of *Curiosity*'s snaps of Mars. The similarities are striking! So, too, for parts of the Gobi Desert (China/Mongolia), the region around Spain's Rio Tinto river, the Hawaiian islands, parts of Tenerife and more. We see what we believe to be similarities in the composition and properties of the rock and soil, the rocky terrain, volcanic formations, geological features, dust-laced air and weather phenomena (dust devils), even glaciers, caves and other habitats that could host extreme life (like those that may be present under the surface of Mars). We even experience Mars-like temperatures in parts of the Norwegian archipelago (Svalbard), northern and eastern Canadian islands, and Antarctica.

Looking at these dusty, arid, red-hued terrestrial surfaces, it is little wonder that we look at Mars and see a world that we could colonise. However, almost all of the locations just mentioned are incredibly difficult for humans to live in comfortably. For example, there is much about Antarctica that we haven't yet explored because of the continent's inhospitable conditions and only a few thousand scientists live there (and not all of those year-round).

However, rising sea levels and global temperatures may make this less of an issue as the years roll past, turning parts of Antarctica into sea and others into thawed terrain that is far more temperate. This is directly relevant for Mars. Rather than an unwanted effect, climate change could actually be used to our advantage if we wished to make an alien environment warmer and more habitable. This is a concept known as terraforming (literally 'Earth-forming' or 'Earth-shaping') or 'planetary ecosynthesis' – deliberately transforming another planet or moon into a more Earth-like environment, and making it more friendly to terrestrial life.

Terraforming has been tackled extensively in science fiction (prominently in Kim Stanley Robinson's *Mars* trilogy) and, although it does have a somewhat scientific basis, it still sits firmly in the sci-fi realm. Terraforming an entire planet would be so prohibitively expensive that it hardly bears considering – especially as it isn't a necessary condition for our colonisation of Mars, and it could take hundreds of thousands of years even if we began today. However, it's quite fun – and scientifically useful – to imagine how we'd go about it.

While the planet's similarities to Earth make it an appealing prospect for potential terraformation (Venus is a contender, too), the process itself seeks to somehow fix the differences. To 'fix' Mars we need to essentially do two things: build an atmosphere around the planet and find a way to keep it there. Such an atmosphere would raise the surface pressure, protect the planet from deep space and – crucially – warm up Mars significantly. This warmth would cause any stores of ice to melt to form bodies of water, recreate and drive a water cycle, and generally create a climate more favourable for terrestrial life.

The main way to warm Mars would be to trigger a greenhouse effect, inflicting human-driven global warming on yet another world. We could do this by releasing substances known for trapping lots of heat from sunlight into the planet's existing atmosphere – ammonia (NH3) has been proposed as a good candidate, as have halocarbons, perfluorocarbons (PFCs) and chlorofluorocarbons (CFCs), some of which could be drawn out from Martian soil and pumped into the air.

Another option is to make Mars's surface darker ('albedo reduction') such that it reflects less sunlight and absorbs more, warming up the planet. This could be done by smashing objects into Mars to unearth deeper and darker material, by breaking Mars's two moons apart so they coat the planet in a thick blanket of dark rubble and dust, or by introducing dark lichens, mosses or algae to its surface, especially in particularly bright areas such as the polar caps.

All of this could raise the temperature from Mars's average -60°C or so (-76°F) up to above zero, and cause most if not all of the carbon dioxide ice locked up at the planet's poles to vaporise, strengthening the warming effect. The water ice at the poles – and elsewhere, if present – would begin to melt, creating surface water and making Mars a far more habitable world. If the melting process needs a little prompting, we could construct huge mirrors and deploy them into orbit around the planet to reflect sunlight down on to Mars's surface – or drop nuclear bombs to really get it moving!

Once we have an atmosphere and surface water, we could then attempt to plant greenery to feed voraciously on the carbon-dioxide-heavy air and pump out oxygen. Watery, warm and covered in plant life, Mars would be well on its way to becoming Earth 2.0.

However, Mars's current atmosphere is far too tenuous to support a greenhouse effect, so we would need to somehow transport a significant amount of material there and induce it to wrap around Mars. If giant orbiting mirrors and pulverised moons weren't enough to get your imagination whirling, this is where the really futuristic suggestions come in.

Some have proposed crumbling Mars's moons into rubble and sending this debris careening downwards to collide with Mars. This would slam into rock and regolith, generate immense heat, and release large amounts of gas in the process (this could be done alongside albedo reduction – two birds, one moon). We could do the same thing by playing cosmic billiards with the asteroid belt, inducing rocky bodies there to hurtle towards Mars. After this, we could introduce huge numbers of radiation-resistant microbes to the Martian soil so they can happily munch away on the nutrients there (water, carbon, nitrates, perchlorates and more), producing puffs of gas (oxygen, ammonia, methane) in the process and, hopefully, reducing the toxicity of the regolith itself. We know this latter technique works, as the oxygen in Earth's atmosphere was pumped out by early microorganisms photosynthesising.

The carbon dioxide content of Martian air currently renders it toxic to humans. It is roughly 95 to 96 per cent carbon dioxide, with a couple of per cent each of argon and nitrogen. Terrestrial air contains just 0.04 per cent carbon dioxide; humans can tolerate a few per cent in the short term, but levels of just 5 per cent would really damage our bodies longer term. We would quickly succumb to carbon dioxide poisoning on Mars in just a breath or two. We need to vastly reduce the carbon dioxide levels – but it's not just a case of adding oxygen, as we would also suffer if the air were pure oxygen. Sadly, Mars has very little in the way of other friendly gases to use as a 'buffer' (primarily nitrogen, which comprises 78 per cent of Earth's atmosphere), so we're not sure of the best way to mix together human-friendly air. NASA is actually working on the idea of generating oxygen from the carbon dioxide in Mars's atmosphere via an experiment named MOXIE (Mars Oxygen ISRU Experiment), to be launched aboard *Mars 2020*.

For this reason, some colonisation attempts suggest small habitats pumped full of breathable air – perhaps a scattering of little domes resembling a bubble-wrap, whack-a-mole society poking up through the Martian dust, underground capsules, or shielded structures (inflatable or permanent) more akin to those in *The Martian*. Our habitats will need to be pressurised, filled with breathable air, spacious enough to avoid Martian 'cabin fever', able to shield us from radiation and relatively simple to erect (we have achieved many of these goals on spacecraft – pressurisation, air, space – and are working on the rest).

However, none of this tackles one of the major differences between Mars and Earth – radiation. Mars has no notable magnetosphere and is constantly bombarded by damaging particles from space, which also strip material from its upper atmosphere. This is 'part two' of the aforementioned terraforming process; we might be able to make an atmosphere, but we need to keep it there. Without a magnetic field, Mars's newly acquired atmosphere would soon be ripped away to space, rendering many of our arduous, expensive and time-consuming efforts useless.

One particularly intriguing terraforming proposal came a few years ago in the form of 'shell worlds', which suggested that we build a giant 'shell' of synthetic Kevlar fibre, steel and dirt around Mars, leaving a small band of space trapped between the shell and the planet's surface. Our communities and cities could then hang from the underside of this shell, with spaceports and other industrial hubs sitting on top like molehills and connected by airlocks.

Safely shielded from the dangers of space radiation, we could then design the inner band of space as we liked – pressures, temperatures, breathable air – and use artificial lighting in place of sunlight.

Even if we built a shell world around Mars, the planet's gravity would still be significantly lower than ours, and microgravity is one of the biggest hurdles to a long-term human presence in space (as described so pessimistically in the previous chapter). One way to alleviate the problems caused by micro- or Martian gravity could well be via some form of artificial gravity. Some missions have suggested 'sky cities' or a 'mothership' in orbit around Mars, to and from which any Martian explorers would periodically shuttle.

We could then simulate Earth-like gravity on such a craft, so astronauts can quickly and periodically return to a familiar environment, by setting one part of the spaceship spinning. This would simulate the tug of artificial gravity and drive anything – and anyone – towards the side of the ship (if the rotating part was a giant wheel, as has been proposed, this would turn the walls of the spacecraft into the 'floor'). It would essentially be a giant centrifuge! We feel a similar kind of effect when making a swift turn in a car, sending passengers hurtling towards one side of the vehicle.

This kind of spinning architecture is often used in films, including *The Martian, Interstellar* and *2001: A Space Odyssey,* presumably because it looks very futuristic and impressive. However, it would make a mission far bulkier and larger than we'd like: the smaller the radius of a spinning ship, the faster it has to spin to generate the same artificial tug of gravity, and the more likely it is to drag the blood within an astronaut's body down towards their feet (leaving them nauseous and ill). Any spaceship that could safely and effectively achieve artificial gravity would need to be significantly bigger than an American football field – and so would also have to be built in space, as there's no way we could launch something that large and massive from the surface of a planet. This may be the Martian system tens of thousands of years from now: spinning motherships orbiting a planet that is slowly being terraformed, with reds and oranges giving way to familiar blues and greens.

Such a mission architecture could do something far more significant than helping us colonise Mars. If we sent humans to Mars and built an increasingly large colony there, the colonisers would begin to slowly adapt to Mars's gravity and find it hard to exist in an Earth-like one (as with astronauts growing in height on the *ISS* and painfully shrinking again when they come back to Earth – if your tendons were stretched for long enough, you would struggle to reverse the effect at all). This is all fine and dandy on a one-way trip, but what if people wish to return one day? It may be that, once there, you're there for the long haul.

Perhaps this is actually something that we should deliberately aim for and it isn't Mars we should change, but ourselves – or so says writer Stephen Petranek. Rather than terraforming Mars, we could 'areform' (Mars-form) humans! Petranek's proposal veers firmly away from science fact and plunges deep into science fiction. He has suggested that we could make colonising Mars easier by genetically modifying our offspring, so that subsequent generations can handle higher levels of carbon dioxide and radiation, lower gravity and surface pressures, thinner air, and could easily metabolise the kinds of plants that could grow well on Mars.

Humankind may splinter to become two distinct groups – humans who can live on Earth and humans who can live on Mars. We could manufacture our very own Martians!

The death of Mars

When it comes to the future of the planet itself, things are relatively bleak.

Regardless of how much humans decide to interfere with the Red Planet, the future of the planets in the Solar System will ultimately be decided by the Sun. Our star seems impressive to us – it contains 99.8 per cent of the mass in our Solar System, is about 4.6 billion years old and a million Earths could comfortably fit inside of it – but it is not all that special on a cosmic scale. It is currently in the 'main sequence' stage of its life, meaning that it is actively fusing hydrogen into helium in its core. The Sun is still less than halfway through its 'adult' life and has over 5 billion years to go before retirement.

Eventually, the Sun will use up all of the hydrogen in its core and will no longer be able to support nuclear fusion, thus beginning its transition to its final stage of life as a planetary nebula. The star will begin to swell up, cooling down as it does so, and become a red giant star. It will then fling its outer layers off into space, completely flooding its surroundings with gas and dust, and leave a small stellar remnant behind (known as a white dwarf star). The radiation from this dwarf will later excite the surrounding material and cause it to glow, as with the bright and beautiful nebulae we spy through our telescopes.

As the Sun puffs its layers into space and evolves, nearby planets will experience fierce winds and shock waves that blast mass and gas from their surfaces and rip it away into space. Smaller planets might lose so much mass that they have very little or none left at all. In its growth to become a red giant the Sun will bloat and expand outwards, effortlessly swallowing up all of the inner planets up to and including Earth – but it might spare Mars*.

---

* Some have suggested that Earth could survive if our planet increases the size of its orbit sufficiently before the destruction begins. Earth's orbit would widen initially as the Sun would lose mass (and thus have a weaker gravity) during its transition to red giant – but the planet would struggle against tidal forces driven by the star and may be sucked inwards. If Earth relocates to somewhere much nearer Mars, it may just about survive. However, the thick winds streaming out from the star may simply blow the entirety of Earth's mass away (or it might collide with Mars).

Even if Mars were able to escape the jaws of a red giant Sun, it would be a completely different world – for many millions of years prior, the rocky residents of our Solar System will be heated and scorched and all of their water and atmosphere ripped away to space by the intense heat, leaving them charred, uninhabitable lumps.

This could all be preceded by far more destructive chaos. Mercury's orbit may destabilise in the next few billion years and wreak havoc in the inner Solar System, driving either itself, Venus or Mars into Earth. Another rogue impactor could streak inwards and hit us in a stroke of very bad luck. A huge burst of energy – a gamma-ray burst, or GRB – could be flung in our direction from far beyond the Solar System. We also know that the Milky Way will merge with the neighbouring Andromeda galaxy in 4 billion years time, completely reshaping our galactic home. Luckily galaxies are mostly empty space so it's likely that the Solar System would survive this collision, but things would look very different.

In the short term, we know for sure that Mars's little family is in for a destructive and turbulent 50 million years or so. Phobos and Deimos will end their lives in completely opposing ways: Phobos, Mars's stretch-marked innermost moon, is slowly spiralling inwards, while the smaller Deimos is moving further away from its parent planet.

Phobos draws a couple of metres closer to Mars every century. This may seem like a trifling amount, but the moon's situation is so fragile that it cannot afford to concede even this much ground without dire consequences. The moon is trapped in a death spiral.

Mars's gravity affects Phobos more and more strongly as the distance between the two bodies lessens. One possible end for Phobos is that the moon will eventually end up getting just too close to Mars and be drawn in to collide with the Red Planet, in an overzealous show of parent-child affection. We think this may have happened before for Mars. The Martian system may have formed with more moons than just Phobos and Deimos, only for these inner siblings to have been dragged downwards and consumed by Mars. Phobos may simply be the next in a long line of infanticidal moonlet murders.

However, models of the Martian system have shown that Phobos will likely soon meet a far more violent end. The clues lie in the moon's red 'stretch marks' – these are thought to be scars from tidal stretching, caused by Mars tugging, twisting, pulling and yanking on Phobos a little too aggressively. Phobos's interior structure is simply too porous and rubble-like to withstand such flexing and reshaping. Rather than colliding with Mars, Phobos may instead be ripped and torn apart while still in orbit, and fragments of the moon flung out into Mars's patch of space to form a ring of rubble. One of the researchers who ran such a model, Benjamin Black of UC Berkeley, US, likened Phobos's demise to 'pulling apart a granola bar, scattering crumbs and chunks everywhere'.

We may not even be able to see Mars's ring from Earth, though it might be possible to see the shadow cast by the ring on the Martian surface. Mars's ring would be composed mostly of dust and rock, the stuff comprising Phobos, whereas the sparkling ring systems of the outer Solar System contain a lot of reflective ice. Although Mars could never rival the beauty of Saturn and its famous ring system, it would still be able to boast of being a ringed planet – for a short period of time, anyway. Scientists estimate that Phobos will meet its sticky end in under 50 – and maybe even as soon as 10 – million years from now. The resulting ring would linger for up to 100 million years before it dissipated from being pulled inwards, hailing down on Mars's surface in rubbly 'moon showers'. The lifespan of the ring depends on how long Phobos is able to withstand Mars's gravity; the closer Phobos gets to Mars before succumbing, the quicker the ring will disappear.

Deimos, meanwhile, is far less attention-seeking than its bigger brother. Rather than making a scene and dying in a cataclysmic explosion, Mars's smallest moon will simply spiral outwards painstakingly slowly, inching further and further from its parent planet.

Our Moon is doing the same thing, only faster; every year it spirals outwards by a few centimetres, and will continue to do so until it has managed to tidally lock our planet – that is, until anyone residing on the Moon would only ever see one side of the Earth, as is the case the other way round. The Earth–Moon pairing will then become a doubly tidally locked system, with the Moon much further away than it is currently. This is to do with the exchange of energy between the two bodies, which slows one down and increases the altitude of the other. Admittedly, humankind will no longer be around to witness this double locking – it would take tens of billions of years for the Moon to tidally lock the Earth – but it is technically possible. Deimos, however, is not nearly massive enough to tidally lock Mars on any timescale, so will drift quietly and unobtrusively out into space.

Given this turbulent future, is it really worth colonising Mars?

For many, despite the fearsome Great Galactic Ghoul and the many challenges of getting there, despite the likelihood of losing vision and muscle mass and mental acuity and bone and even years off one's lifespan, despite missing loved ones and pets and home comforts, despite never again being able to see Earth's night sky, breathe fresh mountain air, sip a favourite beer on a hot summer's day or swim in the sea, despite the arduousness and expense and loneliness and danger and unimaginably long timeline of it all, the answer is still absolutely, unequivocally, 'yes'.

'If I ask any American what happened in 1492, they'll tell me, "Well Columbus sailed in 1492", and that is correct, he did,' said engineer Robert Zubrin, founder and president of the Mars Society and one of the most outspoken proponents of the manned exploration of Mars, in 2014. 'But that's not the only thing that happened in 1492. In 1492, England and France signed a peace treaty. In 1492, the Borgias took over the papacy. In 1492, Lorenzo de' Medici, the richest man in the world, died. A lot of things happened, and if there had been newspapers in 1492 those would have been the headlines, not this Italian weaver's son taking a bunch of ships and sailing off to nowhere. But Columbus is what we remember, not the Borgias taking over the papacy. Well, 500 years from now, people are not going to remember which faction came out on top in Iraq, or Syria, or whatever, and who was in and who was out … but they will remember what we do to make their civilisation possible. This is the most important thing we could do in this time, and if you have it in your power to do something great and important and wonderful, then you should.'

Appendix: Chronological List of Mars Missions

Chronological list of Mars missions 1960–2016 (inclusive)

Launch #	Mission	Year of launch	Type	Agency	Outcome	Still active?
1	*Korabl 4* (*Marsnik 1*)	1960	Fly-by	USSR	Failure	No
2	*Korabl 5* (*Marsnik 2*)	1960	Fly-by	USSR	Failure	No
3	*Korabl 11* (*Sputnik 22*)	1962	Fly-by	USSR	Failure	No
4	*Mars 1* (*Sputnik 23*)	1962	Fly-by	USSR	Failure	No
5	*Korabl 13* (*Sputnik 24*)	1962	Lander	USSR	Failure	No
6	*Mariner 3*	1964	Fly-by	NASA (US)	Failure	No
7	*Mariner 4*	1964	Fly-by	NASA (US)	Success	No
8	*Zond 2*	1964	Fly-by (possible lander)	USSR	Failure	No
9	*Mariner 6*	1969	Fly-by	NASA (US)	Success	No
10	*Mariner 7*	1969	Fly-by	NASA (US)	Success	No
11	*Mars 1969A*	1969	Orbiter	USSR	Failure	No
12	*Mars 1969B*	1969	Orbiter	USSR	Failure	No
13	*Mariner 8*	1971	Orbiter	NASA (US)	Failure	No
14	*Kosmos 419*	1971	Orbiter	USSR	Failure	No
15	*Mars 2*	1971	Orbiter & lander	USSR	Partial success	No
16	*Mars 3*	1971	Orbiter & lander	USSR	Partial success	No
17	*Mariner 9*	1971	Orbiter	NASA (US)	Success	No
18	*Mars 4*	1973	Orbiter	USSR	Failure	No
19	*Mars 5*	1973	Orbiter	USSR	Partial success	No

Launch #	Mission	Year of launch	Type	Agency	Outcome	Still active?
20	*Mars 6*	1973	Fly-by & lander	USSR	Partial success	No
21	*Mars 7*	1973	Fly-by & lander	USSR	Failure	No
22	*Viking 1*	1975	Orbiter & lander	NASA (US)	Success	No
23	*Viking 2*	1975	Orbiter & lander	NASA (US)	Success	No
24	*Phobos 1*	1988	Orbiter & lander	USSR	Failure	No
25	*Phobos 2*	1988	Orbiter & lander(s)	USSR	Partial success	No
26	*Mars Observer*	1992	Orbiter	NASA (US)	Failure	No
27	*Mars Global Surveyor*	1996	Orbiter	NASA (US)	Success	No
28	*Mars 96*	1996	Orbiter & landers (inc. penetrators)	Russian Space Agency (now Roscosmos)	Failure	No
29	*Mars Pathfinder*	1996	Lander	NASA (US)	Success	No
	Sojourner		Rover		Success	No
30	*Nozomi (Planet-B)*	1998	Orbiter	ISAS (now JAXA, Japan)	Failure	No
31	*Mars Climate Orbiter*	1998	Orbiter	NASA (US)	Failure	No
32	*Mars Polar Lander*	1999	Lander	NASA (US)	Failure	No
	Deep Space 2 (Scott and Amundsen)		Landers (penetrators)		Failure	No
33	*2001 Mars Odyssey*	2001	Orbiter	NASA (US)	Success	Yes
34	*Mars Express*	2003	Orbiter	ESA (Europe)	Success	Yes
	Beagle 2		Lander	ESA (UK-led)	Failure	No

Launch #	Mission	Year of launch	Type	Agency	Outcome	Still active?
35	*Mars Exploration Rover A: Spirit*	2003	Rover	NASA (US)	Success	No
36	*Mars Exploration Rover B: Opportunity*	2003	Rover	NASA (US)	Success	Yes
37	*Mars Reconnaissance Orbiter*	2005	Orbiter	NASA (US)	Success	Yes
38	*Phoenix*	2007	Lander	NASA (US)	Success	No
39	*Fobos-Grunt*	2011	Lander (sample return)	Roscosmos (Russia)	Failure	No
	Yinghuo-1		Orbiter	CNSA (China)	Failure	No
40	*Mars Science Laboratory: Curiosity*	2011	Rover	NASA (US)	Success	Yes
41	*Mars Orbiter Mission (Mangalyaan)*	2013	Orbiter	ISRO (India)	Success	Yes
42	*MAVEN (Mars Atmosphere and Volatile EvolutioN Mission)*	2013	Orbiter	NASA (US)	Success	Yes
43	*ExoMars 2016: ExoMars Trace Gas Orbiter*	2016	Orbiter	ESA & Roscosmos (Europe & Russia)	Success	Yes
	ExoMars 2016: Schiaparelli EDM lander		Lander	ESA & Roscosmos (Europe & Russia)	Partial success	No

Acknowledgements

I'd like to say a huge and sincere thank you to everyone who helped me produce this book, including the scientists who patiently fielded my queries and provided their time and expertise, everyone who granted me permission to reprint their wonderful images of Mars, and the fantastic editors and designers at Bloomsbury Sigma for making the process so pleasant. My biggest thank you goes to my ever-supportive husband Angus, who happily and voluntarily took on multiple jobs – proofreader, literary critic, motivational speaker, personal barista – during the writing process.

Index